Nonuniform Line Microstrip Directional Couplers and Filters

For a complete listing of the *Artech House Microwave Library*,
turn to the back of this book

Nonuniform Line Microstrip Directional Couplers and Filters

Sener Uysal

Artech House
Boston • London

Library of Congress Cataloging-in-Publication Data

Uysal, Sener.
Nonuniform Line Microstrip Directional Couplers and Filters/Sener Uysal
Includes bibliographical references and index.
ISBN 0-89006-683-3
1. Microwave circuits—engineering and design 2.Engineering design. II. Title
TK7876.U3892 1993
93-33221
CIP

685 Canton Street
Norwood, MA 02062

International Standard Book Number: 0-89006-683-3
Library of Congress Catalog Card Number: 93-33221

10 9 8 7 6 5 4 3 2 1

To my beautiful wife Aylan
and
Afet (the late) Sermin
Mustafa Gülertan

Contents

Preface

Generally, one finds that the books published in the area of microwave engineering suffer from lack of depth in their treatment of nonuniform line-directional couplers. It is almost impossible to find a book giving the geometrical details of the configurations used. In addition, insufficient examples and the lack of a reliable design procedure have prevented the use of nonuniform microstrip directional couplers, whereas their stripline counterparts have been extensively used in ultrawideband applications. The current book represents the author's efforts to provide microwave engineers with the long-awaited book on nonuniform line microstrip directional couplers and filters.

This book is written primarily for graduate students and research and design engineers. However, parts of the book can be adapted for final-year undergraduate teaching. Development engineers are heavily involved in project management and supervision, which leaves them with very little time to extract design information from detailed analysis and synthesis of nonuniform line directional couplers. Directly applicable graphical data, such as in Chapter 3, and a large number of design examples are provided for the needs of development and contract engineers.

This is the first book to document the real design aspects of nonuniform line directional couplers. Chapters 4 and 5 provide sufficient information for designing cost-effective, high-performance microstrip directional couplers.

The book is the product of the continued work on nonuniform directional couplers that dates back to late 1980s. I would like to thank Charles W. Turner, John Watkins of King's College London, and Gordon Peake of RAE Farnborough, to whom I will always be in debt for their invaluable support and guidance during my time at King's College London. I am grateful to the National University of Singapore for providing support for a project of this scope. Special thanks are due to the anonymous reviewer for the help in putting the book into its final shape. I would also like to thank the editors and staff of Artech House, particularly Julie Lancashire and John Svatek, for their tireless efforts in bringing this book to publication. Last, but not least, I wish to thank my wife, Aylan, for her patience, loving support, suggestions, and skillful typing of the manuscript.

Chapter 1
Introduction

1.1 INTRODUCTORY COMMENTS

Today's military and commercial microwave circuits require wideband systems that can handle numerous data at one time. Directional couplers are the key elements in many wideband microwave circuits such as phase shifters, balanced amplifiers, reflection-type amplifiers, mixers, power dividers, channelizers, modulators, monopulse comparators, and beam-forming networks for array antennas. This book presents a timely contribution to the microwave engineers in this field where the demand for high-performance wideband circuits is continuously increasing.

This book deals with nonuniform line couplers and their applications. Although the book focuses on the planar realization using microstrip technology, it will be evident in later sections that many of the design concepts are equally applicable to other technologies such as stripline.

Ultrawideband performance can be obtained by cascading several couplers together. However, such multisection configurations exhibit unacceptable mismatches due to sharp transitions from low to high coupling and vice versa. Although possible, it is a laborious and tedious task to minimize reflections at those transitions. More often, such modifications also cause performance variations requiring several design iterations. An alternative to multisection coupler design is to employ continuously tapered coupled lines in which sharp transitions can be effectively eliminated.

This book tries to present a balanced combination of theory, design, and application. Several forms of nonuniform line couplers have been clearly distinguished in their operation, capability, and limitations. Detailed derivations are given for many of the equations for design engineers and research students; but those who wish to bypass the details can do so without loss of continuity. Experienced engineers in the field will face no difficulty in extracting the necessary design information for their own system requirements.

The potential problems inherent to nonuniform line directional couplers in inhomogeneous media can be summarized as (1) accurate prediction of performance,

(2) accurate determination of continuous physical parameters, (3) realization of tight coupling values, and (4) isolation. The book presents powerful, cost-effective techniques to solve these problems.

Exact closed-form synthesis functions are presented for accurate performance prediction. Cubic spline interpolation is adapted to synthesize the continuous physical parameters. Tapered interdigitated sections are employed at those sections where the coupling coefficient is too high to be realized by double-coupled lines. Inner edges of double-coupled lines are modified by employing wiggling or serpentining for phase velocity compensation in backward-wave nonuniform line directional couplers. Similar modifications are employed to enhance the difference between the phase velocities in codirectional couplers.

Bandpass coupling is achieved by modifying the reflection coefficient distribution function. Several modification techniques are investigated for subsequent realization of bandpass filters and channelizers. Similar modifications are further used in codirectional couplers to achieve directionality.

It is therefore apparent that of all the advances being made in microwave integrated circuit (MIC) design and technology, some of the most important ones are associated with minimizing tolerance effects and sharpening design procedures. Only in this way will higher grade manufactured circuits result. Clearly there is an urgent need to evolve reliable design documentation for MICs. This book fulfills this need for nonuniform line directional couplers.

1.2 OUTLINE OF THE BOOK

Chapter 1 gives brief analyses for uniform and nonuniform single and coupled microstrip lines. Practical limitations in microstrip directional couplers and filters are discussed in Chapter 2. Several types of directional couplers and filters (in the context of this book, unless otherwise stated, directional couplers or couplers or filters are all assumed to be realized using microstrip technology) are also reviewed in this chapter.

Synthesis, design, and construction of ultrawideband nonuniform directional couplers are given in Chapters 3 and 4. Design data and examples are given for alumina and other high-dielectric-constant substrates. Sensitivity analysis and other practical features that must be considered before implementation are given in Chapter 5. A class of backward-wave couplers that are able to achieve up to almost 0 dB coupling are explained in Chapter 6.

Chapters 7 and 8 describe novel nonuniform coupled-line filters and multiplexers. Codirectional couplers are analyzed in Chapter 9. And finally, some applications of nonuniform directional couplers are discussed in detail in Chapter 10.

The presentation is such that the step-by-step design procedures and examples may be followed by the reader without having to frequently refer to other literature. Selected references are included at the end of each chapter for the benefit of R&D engineers and research students.

1.3 UNIFORM MICROSTRIP LINES

Microstrip is a thin layer of conductor (usually gold or copper) constructed on a dielectric substrate with a ground plane. Its physical construction is shown in Figure 1.1. For many applications the conductor thickness t may be neglected in the analysis. The microstrip medium is referred to as *inhomogeneous*, consisting of the dielectric substrate and air above it. The basic structure of Figure 1.1 is usually modified by shielding or suspending, where necessary to reduce loss, the reflections and parasitic coupling. The worldwide increase in the use of microstrip technology is directly related to its small size, simplicity, reliability, and low cost due to easy manufacturability.

The inhomogeneous media is one of the main problems in the analysis of microstrips. The structure cannot support an exact TEM propagation, and in the resulting dispersion, the conductor dimensions will affect both the characteristic impedance and the velocity of propagation. A quasistatic solution can still be carried out in which the propagation can be approximated to TEM. However, for higher frequencies dispersion becomes significant, and it has to be included in the design of microstrip circuits. Several analysis techniques have evolved for the microstrip problem [1–6]. Some important practical characteristics such as dispersion and loss of microstrip on Barium Tetratitanate substrate ($\epsilon_r = 37$) have been investigated by Lee, Getsinger, and Sparrow [7]. The emerging superconductive microstrip technology has also been analyzed [8]. In this book, the static capacitance analysis will be used. This technique is simple, fast, accurate, and can easily be adapted to coupled-line analysis from which we require individual capacitances to develop a semiempirical design technique for phase velocity compensation in coupled-line directional couplers.

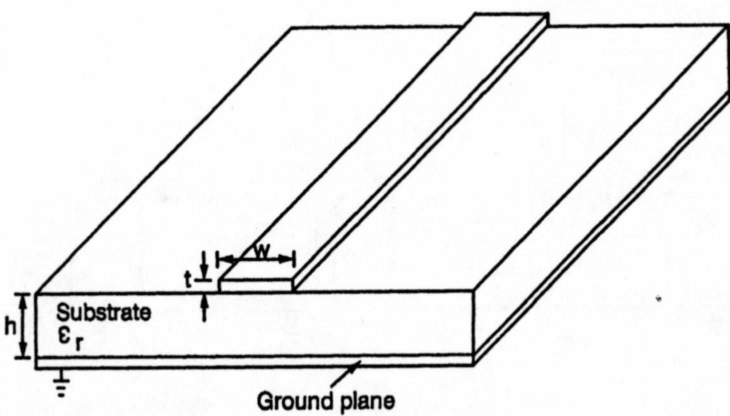

Figure 1.1. Physical construction of microstrip line.

A simplified illustration of the electric field of a microstrip line and its associated static capacitances are given in Figure 1.2. The total capacitance of the microstrip line is given by the sum of the components shown in this figure:

$$C = C_p + 2C_f \tag{1.1}$$

where C_p is the parallel plate capacitance between the conductor and ground plane and C_f is called the fringing capacitance. Closed-form expressions can be obtained for C_p and C_f:

$$C_p = \epsilon_0 \epsilon_r \frac{w}{h} \tag{1.2}$$

$$2C_f = \frac{\sqrt{\epsilon_{re}}}{cZ_{0m}} \tag{1.3}$$

where ϵ_0 is the dielectric constant of air, ϵ_r is the relative dielectric constant of the substrate, ϵ_{re} is the effective dielectric constant, w is the width of the line, h is the substrate thickness, c is the velocity of light in vacuum, and Z_{0m} is the characteristic impedance of the microstrip.

The closed-form expressions for Z_{0m} and ϵ_{re} follow [9]:

$$Z_{0m} = \frac{60}{\sqrt{\epsilon_{re}}} \ln\left(\frac{8h}{w} + 0.25\frac{w}{h}\right) \qquad \left(\frac{w}{h} \leq 1\right) \tag{1.4}$$

$$Z_{0m} = \frac{120\pi}{\sqrt{\epsilon_{re}}} \left\{\frac{w}{h} + 1.393 + 0.667 \ln\left(\frac{w}{h} + 1.444\right)\right\}^{-1} \qquad \left(\frac{w}{h} \geq 1\right) \tag{1.5}$$

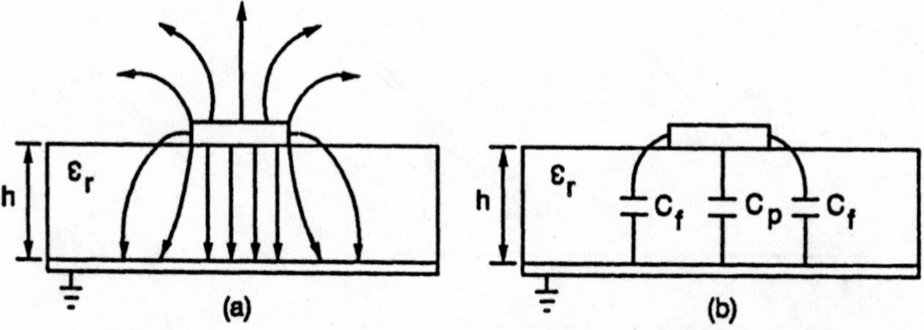

Figure 1.2. (a) Electric field lines for a microstrip line. (b) Static capacitance approximation of the electric field.

$$\epsilon_{re} = \frac{\epsilon_r + 1}{2} + \frac{\epsilon_r - 1}{2} F\left(\frac{w}{h}\right) \tag{1.6}$$

where

$$F\left(\frac{w}{h}\right) = \left(1 + 12\frac{h}{w}\right)^{-1/2} + 0.04\left(1 - \frac{w}{h}\right)^2 \quad \left(\frac{w}{h} \leq 1\right)$$

$$F\left(\frac{w}{h}\right) = \left(1 + 12\frac{h}{w}\right)^{-1/2} \quad \left(\frac{w}{h} \geq 1\right)$$

1.4 UNIFORM COUPLED LINES

A coupled-line configuration consists of two microstrip lines parallel to each other and in close proximity. The electromagnetic wave traveling on one line induces a current on the adjacent line.

Due to the coupling of electromagnetic fields, the coupled lines can support two different modes of propagation with different characteristic impedances. The velocity of propagation of these two modes is equal when the lines are imbedded in a homogeneous dielectric medium. However, for coupled microstrip lines the dielectric medium is not homogeneous. A part of the field extends into the air above the substrate. This fraction is different for the two modes of coupled lines. Consequently, the effective dielectric constants (and the phase velocities) are not equal for the two modes. This feature of coupled microstrip lines becomes highly undesirable when they are used in backward-wave directional couplers, edge-coupled line filters, and several other circuits. Solution to this problem will be given in later chapters.

Among several other methods, this book concentrates on the quasistatic even- and odd-mode capacitance analysis of coupled lines. This technique is simple, accurate, and can be used in the derivation of semiempirical closed-form equations for phase velocity compensation. A general case of N-element coupled-line configuration is shown in Figure 1.3. In the even-mode excitation these lines are excited with signals of equal amplitude and phase. Therefore, a magnetic wall can be placed between the

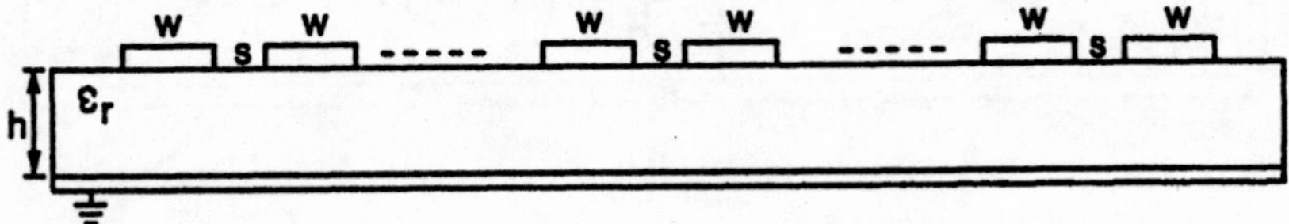

Figure 1.3. Uniform N-element coupled microstrip lines.

lines. In the odd-mode the alternating pairs of coupled lines are excited with signals of equal amplitude having opposite phases. Hence, an electric wall can be placed between the lines. The static capacitance parameters for these two modes are shown in Figure 1.4. Closed-form expressions for the capacitances for the even- and odd-mode have been derived in [10] and will follow.

The fringing capacitance modified by the presence of a second conductor is given by

$$C_{fe} = \frac{C_f}{1 + A(h/s)\tanh(10s/h)} \sqrt{\frac{\epsilon_r}{\epsilon_{re}}} \tag{1.7}$$

where $A = \exp[-0.1 \exp(2.33 - 2.53w/h)]$ and C_f is given by equation (1.3).

The capacitances between the conductors are given by

$$2C_{gd} = \frac{\epsilon_0 \epsilon_r}{\pi} \ln\left[\coth\left(\frac{\pi s}{4h}\right)\right] + 0.65 C_f \left\{0.02 \frac{h}{s} \sqrt{\epsilon_r} + \left(1 - \frac{1}{\epsilon_r^2}\right)\right\} \tag{1.8}$$

$$2C_{ga} = \frac{\epsilon_0}{\pi} \ln\left\{2 \frac{1 + \sqrt{k'}}{1 - \sqrt{k'}}\right\} \qquad 0 \le k^2 \le 0.5 \tag{1.9}$$

$$2C_{ga} = \pi \epsilon_0 \left[\ln\left(2 \frac{1 + \sqrt{k}}{1 - \sqrt{k}}\right)\right]^{-1} \qquad 0.5 \le k^2 \le 1 \tag{1.10}$$

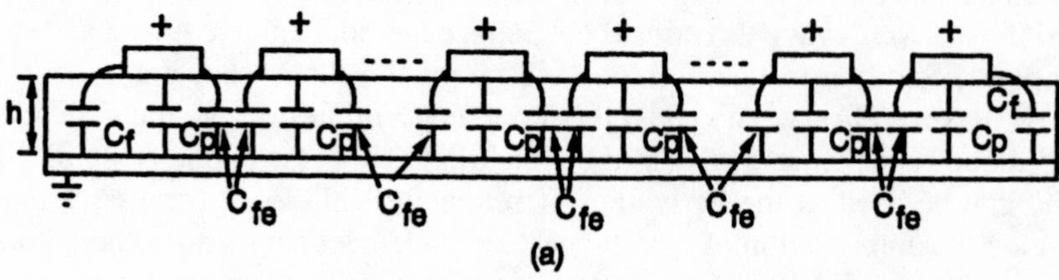

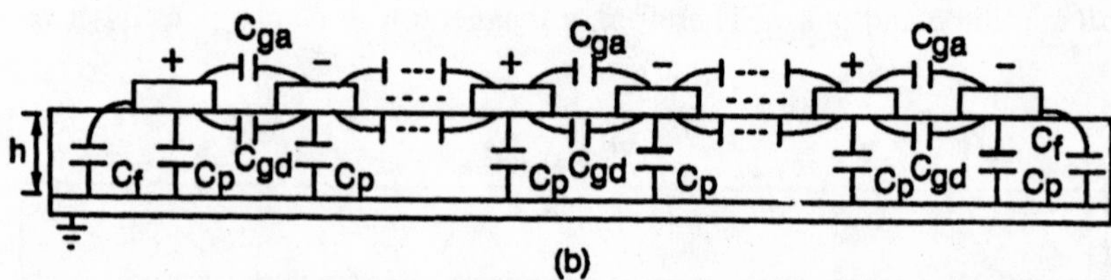

Figure 1.4. Static capacitances for N-element coupled microstrip lines: (a) even-mode and (b) odd-mode.

$\coth z = \frac{e^z + e^{-z}}{e^z - e^{-z}}$

where

$$k = \frac{s/h}{s/h + 2w/h}$$

$$k' = \sqrt{1 - k^2}$$

The even- and odd-mode capacitances for the general case with N-coupled lines can be expressed in terms of the individual capacitances as follows:

$$C_e = C_f + \left(\frac{N}{2}\right) C_p + (N - 1)C_{fe} \tag{1.11}$$

$$C_o = C_f + \left(\frac{N}{2}\right) C_p + (N - 1)C_{fo} \tag{1.12}$$

where $C_{fo} = C_{ga} + C_{gd}$. Cross coupling between alternating elements is neglected.

Next we consider the even- and odd-mode capacitance analysis reported by Smith [11]. Smith uses a conformal mapping technique to obtain the static capacitance parameters for the general configuration shown in Figure 1.1. His equations will not be given here; the reader may refer to [11] for a proper understanding of his method.

For either method described proviously, the corresponding characteristic impedances for the two modes can be expressed in terms of the static capacitances:

$$Z_{0e} = \frac{1}{c\sqrt{C_e C_{ea}}} \qquad Z_{0o} = \frac{1}{c\sqrt{C_o C_{oa}}} \tag{1.13}$$

where c is the velocity of light in vacuum and C_{ea} and C_{oa} denote the even- and odd-mode capacitances with dielectric replaced by air.

The coupling coefficient k and the characteristic impedance Z_0 can be defined in terms of even- and odd-mode impedances as

$$k = \frac{Z_{0e} - Z_{0o}}{Z_{0e} + Z_{0o}} \tag{1.14}$$

$$Z_0 = \sqrt{Z_{0e} Z_{0o}} \tag{1.15}$$

The effective dielectric constants and the phase velocities for the two modes are given by [10]

$$\epsilon_{re} = \frac{C_e}{C_{ea}} \qquad \epsilon_{ro} = \frac{C_o}{C_{oa}} \tag{1.16}$$

We now introduce subscripts 1 and 2 for the equations of Garg and Bahl [10] and Smith [11], respectively; then

$$C_{fm} = \frac{C_{f1} + C_{f2}}{2} \qquad C_{fem} = \frac{C_{fe1} + C_{fe2}}{2}$$

$$C_{fom} = \frac{C_{fo1} + C_{fo2}}{2} \qquad C_{em} = \frac{C_{e1} + C_{e2}}{2}$$

$$C_{om} = \frac{C_{o1} + C_{o2}}{2} \tag{1.17}$$

$$C_{fam} = \frac{C_{fa1} + C_{fa2}}{2} \qquad C_{feam} = \frac{C_{fea1} + C_{fea2}}{2}$$

$$C_{foam} = \frac{C_{foa1} + C_{foa2}}{2} \qquad C_{eam} = \frac{C_{ea1} + C_{ea2}}{2}$$

$$C_{oam} = \frac{C_{oa1} + C_{oa2}}{2}$$

where m denotes average value. We can rewrite equation (1.13) as

$$Z_{0e}Z_{0o}c^2\sqrt{C_eC_{ea}C_oC_{oa}} - 1 = 0 \tag{1.18}$$

or using equation (1.15), we obtain

$$Z_0^2c^2\sqrt{C_eC_{ea}C_oC_{oa}} - 1 = 0 \tag{1.19}$$

This equation is valid for zero frequency only. Frequency dependence can be introduced as follows:

$$Z_0^2c^2C_{ea}C_{oa}\sqrt{\epsilon_{re}(f)\epsilon_{ro}(f)} - 1 = 0 \tag{1.20}$$

The values of $\epsilon_{re}(f)$ and $\epsilon_{ro}(f)$ can be obtained by using Getsinger's [12] formulation. However, for a better accuracy the frequency-dependent closed-form equations of Hammerstad and Jensen [13] or Kirschning and Jansen [14] can be employed.

Equation (1.20) forms the basis of our optimization. This equation is a function of Z_0, ϵ_r, h, w, s, and f. Because Z_0, h, ϵ_r, and f are known design parameters and, for a given Z_0 only one pair of shape ratios (w/h, s/h) satisfy the given impedance and the amount of coupling required simultaneously, (1.20) can be optimized for a given value of s to yield the corresponding value of w. This equation is optimized for s values ranging from 0.05 to 2.5 mm by using Garg and Bahl [10], Smith [11], and their average values. A characteristic impedance of 50 Ω was specified on 0.635-mm-thick alumina substrate with $\epsilon_r = 9.9$.

The individual even- and odd-mode capacitances corresponding to this optimization are plotted as a function of coupling coefficient k and are given in Figure 1.5. The parallel plate capacitance C_p is included in Figure 1.5(a) for completeness. In Figure 1.5(b) C_{fo} is plotted up to $k = 0.625$ because C_{fo1} was increasing sharply.

We can see from this figure that the capacitance values reported by Garg and Bahl [10] and Smith [11] have an increasing difference for $k > 0.6$. We then check the accuracy of their average values. This is done by comparing the corresponding optimized physical dimensions w and s against Touchstone™ values for the same amount of coupling coefficient. We find that the difference is less than 0.1% for $N = 2$ and 0.35% for $N = 4$. The accuracy of this technique has been repeatedly tested against other available data for a wide range of dielectric constant values. In all cases, the accuracy proved satisfactory. We shall refer to this section in Chapter 3. We need uniform coupled-line data, such as phase velocities, to synthesize nonuniform coupled line performance. Also we need the physical dimensions of uniform coupled lines to find continuous physical dimensions for nonuniform coupled lines. How we determine these will be explained in Chapter 3.

1.5 NONUNIFORM LINES

Nonuniform transmission lines [15–18] are widely used as broadband impedance-matching elements. A section of such a line is shown in Figure 1.6. The general transmission line equations are

$$\frac{dV(x)}{dx} = -Z(x)I(x) \tag{1.21}$$

$$\frac{dI(x)}{dx} = -Y(x)V(x) \tag{1.22}$$

where $V(x)$ is the voltage across the line, $I(x)$ is the current in the line, and $Z(x)$ and $Y(x)$ are the series impedance and shunt admittance per unit length of the line.

The propagation constant, the characteristic impedance, and the reflection coefficient are given by the following relationships:

$$\gamma(x) = \sqrt{Z(x)Y(x)} \tag{1.23}$$

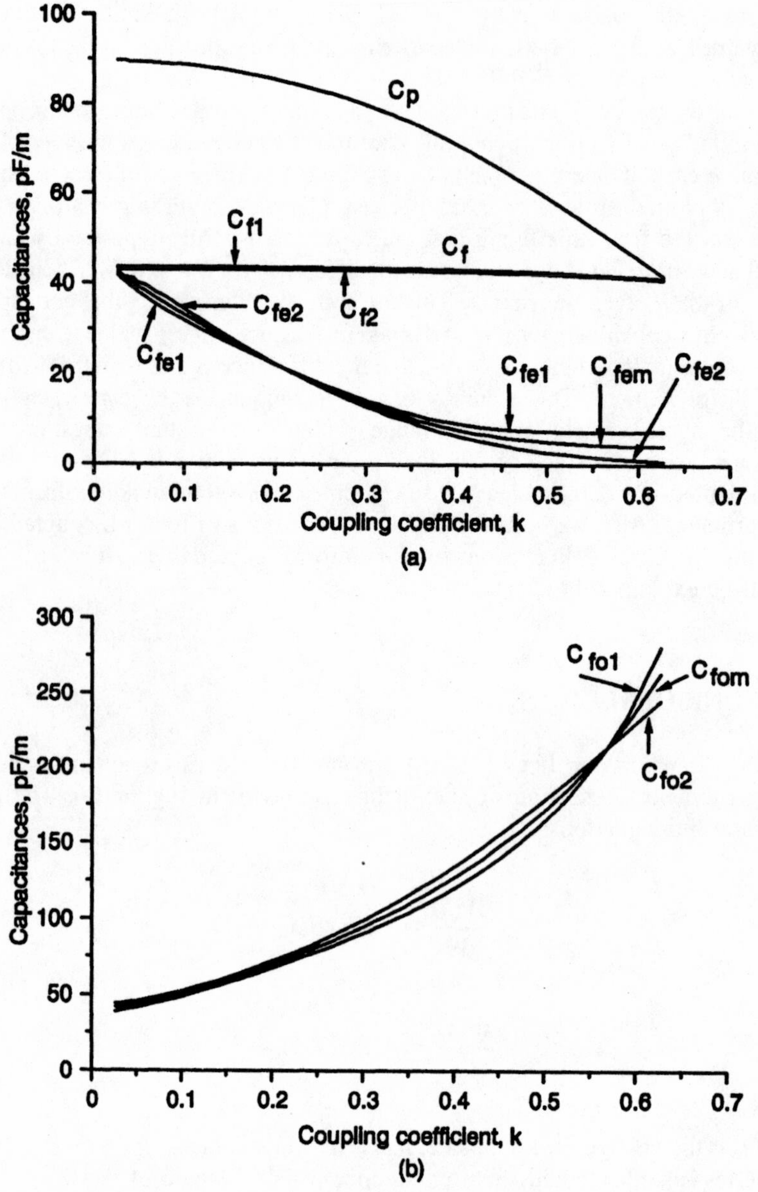

Figure 1.5. Optimized capacitances for 50-Ω couplers on alumina substrate with $\epsilon_r = 9.9$, $h = 0.635$ mm, and $N = 2$: (a) even-mode capacitance C_{fe} and (b) odd-mode capacitance C_{fo}.

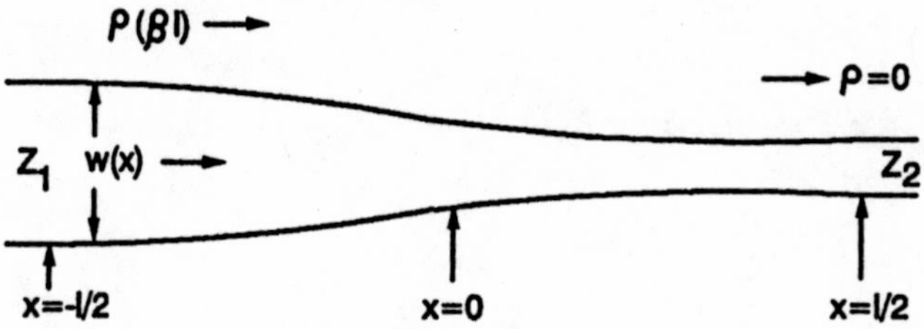

Figure 1.6. Nonuniform microstrip line section.

$$Z_0(x) = \sqrt{\frac{Z(x)}{Y(x)}} \tag{1.24}$$

$$\rho(x) = \frac{V(x)/I(x) - Z_0(x)}{V(x)/I(x) + Z_0(x)} \tag{1.25}$$

Differentiating eq. (1.25) with respect to x and using eqs. (1.23)–(1.25) we obtain a first-order nonlinear differential equation as

$$\frac{d\rho}{dx} - 2\gamma\rho + (1 - \rho^2)p(x) = 0 \tag{1.26}$$

where

$$p(x) = \frac{1}{2}\frac{d}{dx}\ln Z_0(x) \tag{1.27}$$

where $Z_0(x)$ is the normalized characteristic impedance.

Equation (1.26) is a first-order nonlinear differential equation that can be solved for ρ by using a series solution reported by Bergquist [15]. His equations are summarized here:

$$\rho(x) = \frac{\phi_1 + \rho(0)\psi_2}{\phi_2 + \rho(0)\psi_1}\exp\left(2\int_0^x \gamma(x)\,dx\right) \tag{1.28}$$

where

$$\phi_1 = K_1 + K_3 + K_5 + \cdots$$
$$\phi_2 = 1 + K_2 + K_4 + \cdots$$
$$\psi_1 = Q_1 + Q_3 + Q_5 + \cdots$$
$$\psi_2 = 1 + Q_2 + Q_4 + \cdots$$

$$K_1 = \int_0^x f_1(x)\,\mathrm{d}x$$

$$K_2 = \int_0^x f_2(x)K_1\,\mathrm{d}x$$

$$K_3 = \int_0^x f_1(x)K_2\,\mathrm{d}x$$

$$K_4 = \int_0^x f_2(x)K_3\,\mathrm{d}x$$

$$\vdots$$

$$Q_1 = \int_0^x f_2(x)\,\mathrm{d}x$$

$$Q_2 = \int_0^x f_1(x)Q_1\,\mathrm{d}x$$

$$Q_3 = \int_0^x f_2(x)Q_2\,\mathrm{d}x$$

$$Q_4 = \int_0^x f_1(x)Q_3\,\mathrm{d}x$$

$$\vdots$$

$$f_1(x) = p(x)\exp(-2\int_0^x \gamma(x)\,\mathrm{d}x)$$

$$f_2(x) = p(x)\exp(2\int_0^x \gamma(x)\,\mathrm{d}x)$$

1.6 NONUNIFORM COUPLED LINES

The coupled-arm response of nonuniform coupled lines can be synthesized with the aid of nonuniform transmission lines [19]. A nonuniform transmission line is assumed to have a characteristic impedance curve equal to the even-mode impedance curve of the coupled lines to be analyzed. Under these conditions the reflection coefficient of the transmission line is equal in magnitude and phase to the coupled-arm response of the coupler. The series solution for the reflection coefficient given by eq. (1.28) can be used as the coupling response for the nonuniform coupler shown in Figure 1.7.

In our case, we shall assume that the ports are perfectly matched. That is, we set $\rho(0) = 0$ in eq. (1.28). We then obtain

$$\phi_1 = G + G^3/3! + G^5/5! + \cdots = \sinh(G)$$
$$\phi_2 = 1 + G^2/2! + G^4/4! + G^6/6! + \cdots = \cosh(G)$$

where $G = |\int_0^l \sin(2\omega v/v)p(x)\,dx|$.

Since we have $\rho(x)$ (single line) $\equiv C(\omega)$ (coupled line), we then have the coupling function as [20]

$$C(\omega) = \tanh\left\{\int_0^l \sin(2\omega x/v)p(x)\,dx\right\} \tag{1.29}$$

where l is the overall coupler length and v is the velocity of propagation in the medium and $p(x)$ is given by

$$p(x) = \frac{1}{2}\frac{d}{dx}\ln Z_{0e}(x) \tag{1.30}$$

where $Z_{0e}(x)$ is the normalized even-mode impedance. The quantity $p(x)$ is known as the *reflection coefficient distribution function* and forms a Fourier transform pair with

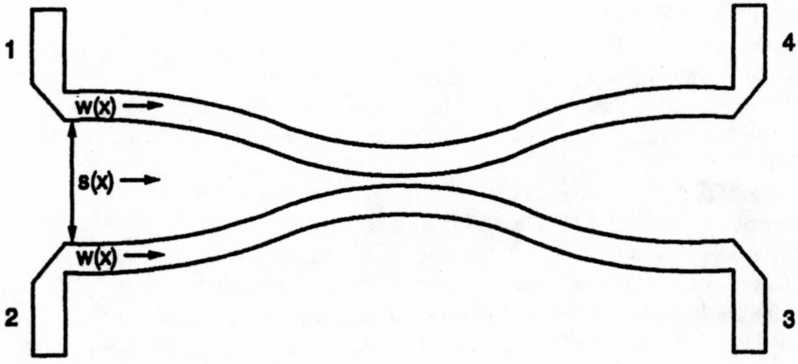

Figure 1.7. Nonuniform double-coupled lines.

eq. (1.29) [21]; therefore $p(x)$ can be obtained from equation (1.29) by an inverse Fourier transformation:

$$p(x) = -\frac{2}{\pi v}\int_0^{2\omega_c} \sin\left(\frac{2\omega x}{v}\right) \tanh^{-1}[C(\omega)]\, d\omega \tag{1.31}$$

where ω_c is the design center frequency. In these derivations we have neglected the losses and assumed a constant velocity of propagation in the guide.

REFERENCES

[1] Yamashita, E., and R. Mittra, "Variational Method for the Analysis of Microstrip Lines," *IEEE Trans. on Microwave Theory and Tech.,* Vol. MTT-16, 1968, pp. 251–256.

[2] Daly, P., "Hybrid-Mode Analysis of Microstrip by Finite Element Methods," *IEEE Trans. on Microwave Theory and Tech.,* Vol. MTT-19, 1971, pp. 19–25.

[3] Stinehelfer, H. E., "An Accurate Calculation of Uniform microstrip Transmission Lines," *IEEE Trans. on Microwave Theory and Tech.,* Vol. MTT-16, 1968, pp. 439–444.

[4] Krage, M. K., and G. I. Haddad, "Frequency-Dependent Characteristics of Microstrip Transmission Lines," *IEEE Trans. on Microwave Theory and Tech.,* Vol. MTT-20, 1972, pp. 678–688.

[5] Schneider, M. V., "Microstrip Lines for Microwave Integrated Circuits," *Bell System Tech. J.,* May–June 1969, pp. 1421–1444.

[6] Getsinger, W. J., "Measurement and Modelling of Apparent Characteristic Impedance of Microstrip," *IEEE Trans. on Microwave Theory and Tech.,* Vol. MTT-31, 1983, pp. 624–632.

[7] Lee, Y.S., W. J. Getsinger, and L. R. Sparrow, "Barium Tetratitanate MIC Technology," *IEEE Trans. on Microwave Theory and Tech.,* Vol. MTT-27, 1979, pp. 655–660.

[8] Ganguly, A. K., and C. M. Krowne, "Characteristics of Microstrip Transmission Lines with High-Dielectric-Constant Substrates," *IEEE Trans. on Microwave Theory and Tech.,* Vol. MTT-39, 1991, pp. 1329–1337.

[9] Hammerstad, E. O., "Equations for Microstrip Circuit Design," *Proc. European Microwave Conf.,* 1975, pp. 268–272.

[10] Garg, R., and I. J. Bahl, "Characteristics of Coupled Microstriplines," *IEEE Trans. on Microwave Theory and Tech.,* Vol. MTT-27, 1979, pp. 700–705.

[11] Smith, J. I., "The Even- and Odd-Mode Capacitance Parameters for Coupled Lines in Suspended Substrate," *IEEE Trans. on Microwave Theory and Tech.,* Vol. MTT-19, 1971, pp. 424–431.

[12] Getsinger, W. J., "Dispersion of Parallel-Coupled Microstrip," *IEEE Trans. on Microwave Theory and Tech.,* Vol. MTT-21, 1973, pp. 144–145.

[13] Hammerstad, E., and Ø. Jensen, "Accurate Models for Microstrip Computer-Aided Design," *IEEE MTT-S Int. Microwave Symp. Digest,* 1980, pp. 407–409.

[14] Kirschning, M., and R. H. Jansen, "Accurate Wide-Range Design Equations for the Frequency-Dependent Characteristic of Parallel Coupled Microstrip Lines," *IEEE Trans. on Microwave Theory and Tech.,* Vol. MTT-32, 1984, pp. 83–90.

[15] Bergquist, A., ''Wave Propagation on Nonuniform Transmission Lines,'' *IEEE Trans. on Microwave Theory and Tech.,* Vol. MTT-20, 1972, pp. 557–558.

[16] Klopfenstein, R. W., ''A Transmission Line Taper of Improved Design,'' *Proc. I.R.E.,* Vol. 44, 1956, pp. 31–35.

[17] Youla, D. C., ''Analysis and Synthesis of Arbitrarily Terminated Lossless Nonuniform Lines,'' *IEEE Trans. on Circuit Theory,* 1964, pp. 363–372.

[18] Protonotarios, E. N., and O. Wing, ''Analysis and Intrinsic Properties of the General Nonuniform Transmission Line,'' *IEEE Trans. on Microwave Theory and Tech.,* Vol. MTT-15, 1967, pp. 142–150.

[19] Tresselt, C. P., ''The Design and Construction of Broadband High-Directivity, 90-Degree Couplers Using Nonuniform Line techniques,'' *IEEE Trans. on Microwave Theory and Tech.,* Vol. MTT-14, 1966, pp. 647–656.

[20] Uysal, S., and J. Watkins, ''Novel Microstrip Multifunction Directional Couplers and Filters for Microwave and Millimeter-Wave Applications,'' *IEEE Trans. on Microwave Theory and Tech.,* Vol. MTT-39, 1991, pp. 977–985.

[21] Kammler, D. W., ''The Design of Discrete *N*-Section and Continuously Tapered Symmetrical Microwave TEM Directional Couplers,'' *IEEE Trans. on Microwave Theory and Tech.,* Vol. MTT-17, 1969, pp. 577–590.

Chapter 2
Review of Microstrip Directional Couplers and Filters

2.1 INTRODUCTION

The manufacture of microstrip circuits has now reached a mature stage, with well-defined manufacturing guidelines for both prototype development at the laboratory and mass production. Although the latter may involve automated production, which can yield repeatable high-grade circuits, it is not completely free from certain limitations, brought about by certain characteristics of microstrip circuits. These limitations become more clear when large-scale integration is required.

With microstrip directional couplers, spacing between conductors becomes a critical factor in controlling the amount of coupling to the adjacent microstrip conductor. A small deviation from the required separation, especially when tight coupling is required, may result in unacceptable circuit performance. Similar limitations also exist for certain types of microstrip filters. In the following sections we shall attempt to identify major practical limitations associated mainly with microstrip coupled-line circuits. A brief overview of microstrip directional couplers and filters will also be given. It will then be evident that certain practical limitations have actually benefited microwave engineers, including the author of this book. Novel circuits, whose details will be given in Chapters 6, 7, and 8, were invented while trying to overcome certain limitations in coupled lines. Therefore, it is of paramount importance to have an in-depth understanding of these problems associated with coupled microstrip lines.

2.2 PRACTICAL LIMITATIONS IN MICROSTRIP DIRECTIONAL COUPLERS

Practical limitations in microstrip directional couplers can be given under three main categories: losses, physical dimensions, and directivity. Ohmic loss, dielectric loss

[1,2], and radiation loss [3] are the main losses in microstrip couplers. Transition loss (coaxial-to-microstrip and microstrip-to-coaxial) [4], which is encountered at component-level design, will be omitted from this list because it is not intrinsic to the circuit. Thus far, partial solutions have been introduced to minimize these losses. Radiation loss can be kept to a minimum (if not completely eliminated) by shielding the circuit. Ohmic loss can be reduced by employing good conductors, such as copper and gold. This loss can also be reduced by employing superconductive materials. However, the technique is not yet mature and presently is too expensive for use with every microstrip circuit. On the other hand, superconductivity is very attractive for systems requiring large volume of microstrip components and where available power is scarce (e.g., space applications). A technology akin to microstrip, known as *suspended substrate,* has also been widely employed because of its ability to reduce both radiation and ohmic losses. It does so because the microstrip circuitry on the substrate is placed between two layers of grounded substrates that act as a shield to radiation. This results in increased conductor width for the same characteristic impedance of microstrip. Although such a solution is accompanied by an increase in production cost and weight, it is widely used because solid-state device circuit integration is straightforward as in microstrip. Similar solutions, also in the form of modifying the basic microstrip circuitry, have been widely investigated [5,6] but will not be included in this book, as different applications may have different requirements.

Any reduction in dielectric loss simply follows the development of substrate technology. It is simply not possible to fabricate every circuit designed on paper; in many cases, physical dimensions can be prohibitive. Although there exist very accurate fabrication methods, these are not available to everyone (especially to those researchers working in a university laboratory). Yet for a given substrate thickness and dielectric constant there is a limit on conductor width onset by the realizable characteristic impedance. Very wide lines, even if shielded, may cause excitation of spurious modes, especially at higher frequencies. Yet again very high impedance lines may become so narrow that they may fall off the substrate. In addition to these, physical dimensions become a major problem when power handling capability and component lifetime are taken into consideration.

Another practical limitation imposed on the physical dimensions is the dispersive nature of the microstrip. As it will be seen in later chapters, this frequency dependence may differ significantly for the even and odd modes for microstrip-coupled lines. This may result in performance degradation in wideband couplers due to deterioration in directivity, which will be discussed next.

Directivity in directional couplers can be defined as the absolute value of $20 \log(S_{31}/S_{21})$. It is a measure of coupler quality, and ideally its value is infinite when no signals exit from the isolated port. Thus far directivity has been the most important parameter in directional coupler design. The scattering parameters, namely, S_{11}, S_{21}, S_{31}, and S_{41}, are all functions of even- and odd-mode phase velocities. The difference between the phase velocities, therefore, affects all four parameters. Except in those applications where all four ports are utilized, the isolated port is usually terminated in the characteristic impedance of the design to minimize interstage reflections from it. However, terminating this port does not always provide a satisfactory solution for those applications where both amplitude and phase quadrature properties of the coupler are required (e.g., -3 dB coupler used as a power-divider circuit for the *I* and *Q* channels of a modulator). Therefore, it is essential to introduce some form of modification for phase velocity equalization. On the other hand, the design principles applicable to bandpass-type couplers can be used to design forward-wave couplers with excellent directivity without having to use any modification for phase velocity compensation. This will be discussed in Chapter 9.

2.3 PRACTICAL LIMITATIONS IN MICROSTRIP FILTERS

The limitations imposed on directional couplers by losses and physical dimensions are equally applicable to microstrip filters. In addition to these limitations microstrip filters suffer from the lack of a generally applicable design procedure. Very narrow bandwidths, which are normally required for communication channels at microwave frequencies, are extremely difficult to achieve. Several cascaded resonator sections may be required for sharp cutoff and high out-of-band rejection. Any such performance improvement is accompanied by an increased loss in the filter passband. Superconductive microstrip filters may be an alternative in obtaining high-Q microstrip filters. Another recognized limitation in microstrip filter design is again the dispersive nature of microstrip, which may cause a significant shift in the design center frequency.

2.4 TYPES OF MICROSTRIP COUPLERS

Microstrip directional couplers can be grouped under two main categories: (1) backward-wave couplers and (2) forward-wave couplers. Each category can be further classified as symmetrical or asymmetrical and asymmetry or symmetry can be dimensional or axial. A classification tree for microstrip directional couplers is given in Figure 2.1(a). In the classification tree, it is assumed that the number of coupled sections is greater than 1 (in the case of nonuniform couplers, the coupler length at midband frequency is greater than ($\lambda_c/4$). Otherwise, the terms *axial symmetry* and *asymmetry* are not valid. Figure 2.1 (b) and (c) provide the coupler notation for the backward-wave and forward-wave couplers that are used throughout the book.

The simplest form of microstrip directional coupler [7–11] is given in Figure 2.2. This coupler consists of two identical lines of width w and separated by a uniform gap s. For a given substrate thickness, h, the physical dimensions are expressed as shape ratios w/h and s/h. Coupling from one line to the other depends critically on the spacing between the lines. The width affects the impedance level of the system and is not very

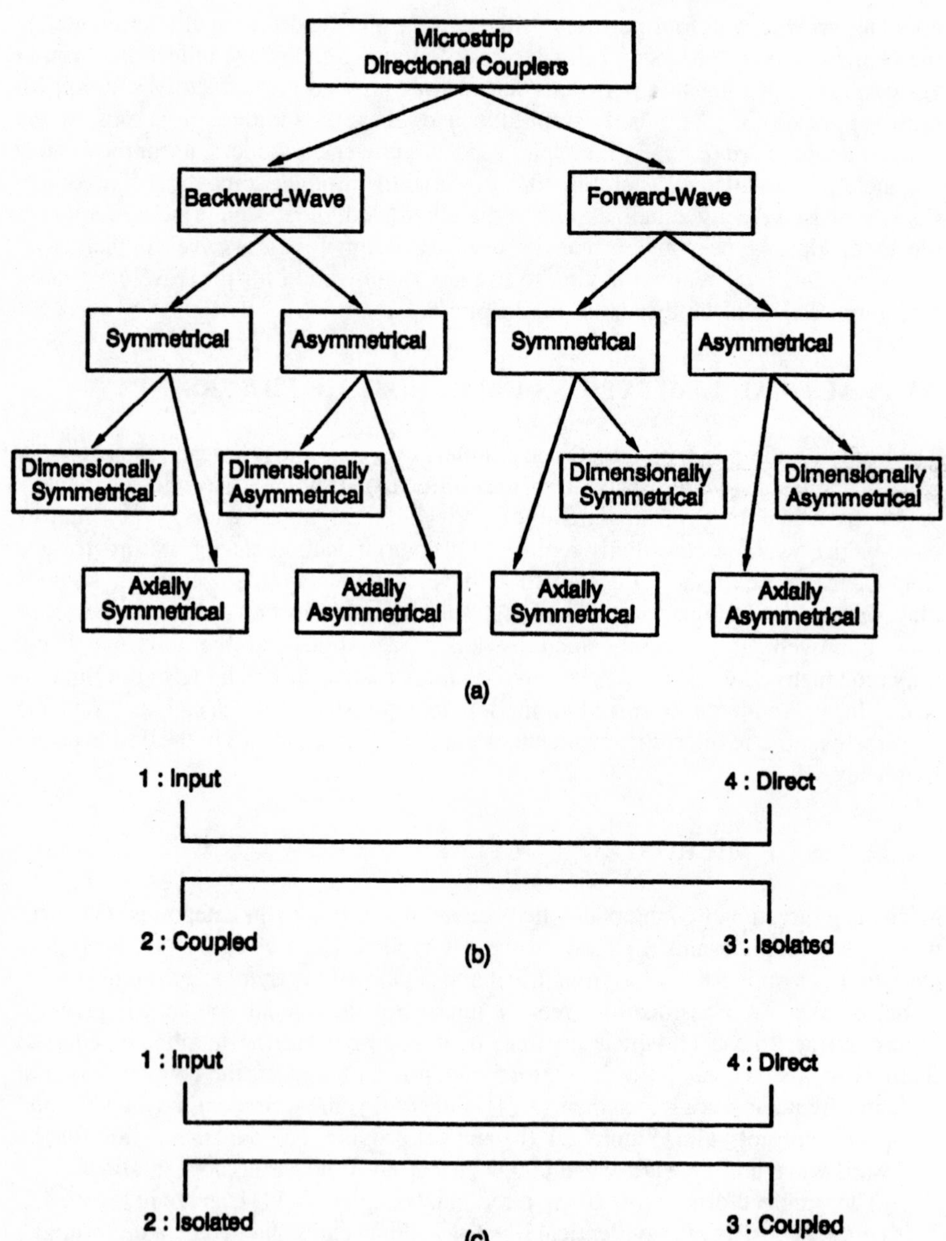

Figure 2.1. (a) Classification tree for microstrip directional couplers. (b) Notation for backward-wave couplers. (c) Notation for forward-wave couplers.

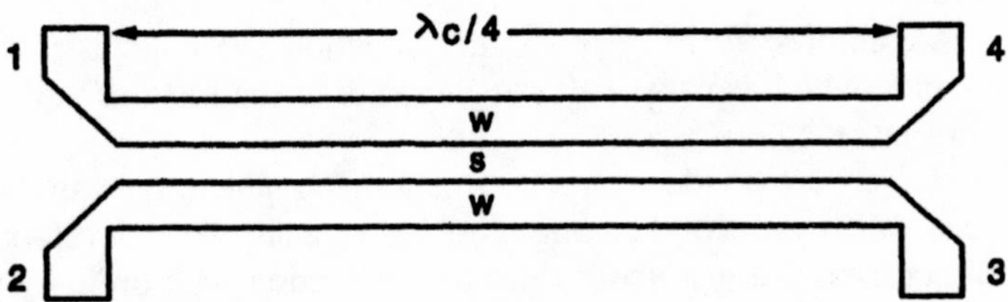

Figure 2.2. Uniform double-coupled coupler configuration.

sensitive to the amount of coupling. The coupler is a quarter-wavelength long at the midband. It is customary to use optimum mitered bends for, at least, two of the ports (one at each end) to minimize reflections from these bends [12]. This coupler is usually employed in those applications where low coupling is required. It is also widely used as the main building block for edge-coupled line filters. The maximum coupling achievable with this coupler using ordinary photolithographic techniques lies somewhere between -6 and -8 dB depending on the dielectric constant and thickness of the substrate.

A three-line coupler, shown in Figure 2.3, can be used to achieve moderate coupling [13]. In this case we have two different widths, w_1 and w_2, and the lines are separated by uniform gaps, denoted by s. Because we have two gaps, it is essential to make a physical connection between first and third lines to suppress spurious modes. The existence of two gaps increases the odd-mode capacitance, thereby increasing the amount of coupling. Ideally, the bond wires [14] usually used (ribbon may also be used) are required to behave like short circuits. Therefore, to reduce inductive effects at those crossovers, more than one, and usually three, bond wire is used.

The most common and most useful type of directional coupler is a -3 dB quadrature coupler. Because the input power is divided equally into two ports, the quadrature phase characteristic between these ports becomes very useful in numerous

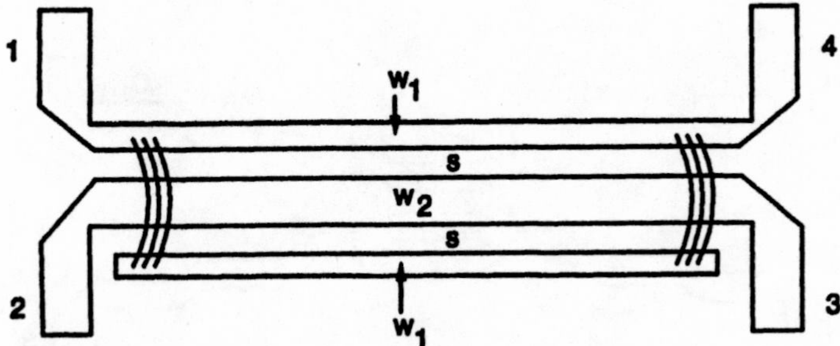

Figure 2.3. Uniform three-line coupler.

applications. The simplest form of a −3 dB directional coupler is a four-line interdigitated coupler, which is named after its inventor as the Lange coupler [15]. This is illustrated in Figure 2.4.

Alternate fingers are bonded to reduce the four possible modes to two. Because of its importance, other forms of the Lange coupler have also been developed [16–21]. Its physical dimensions are just about right for realization with ordinary photolithographic techniques. Its small size makes it a preferred component with monolithic microwave integrated circuit (MMIC) applications. Although, it can give an octave band performance it is not completely free from limitations. Bond wires may cause performance variations and therefore may limit its use to applications below 30 GHz. However, GaAs monolithic versions may be used up to 100 GHz.

Ultrawideband performance can be obtained by cascading several couplers together, as illustrated in Figure 2.5. This coupler consists of N coupled elements (where N is an odd number greater than 1 and equal to $2n - 1$), each of which is a quarter-wavelength long at the design center frequency. The tightest coupled section is at the middle with identical left and right coupled sections. The transitions from low-to-high coupling and vice versa present mismatches to both direct and coupled signals. Therefore, even with compensated transitions, the number of cascadable sections is limited, and this type of coupler is suitable only for low coupling.

Figure 2.4. Lange coupler.

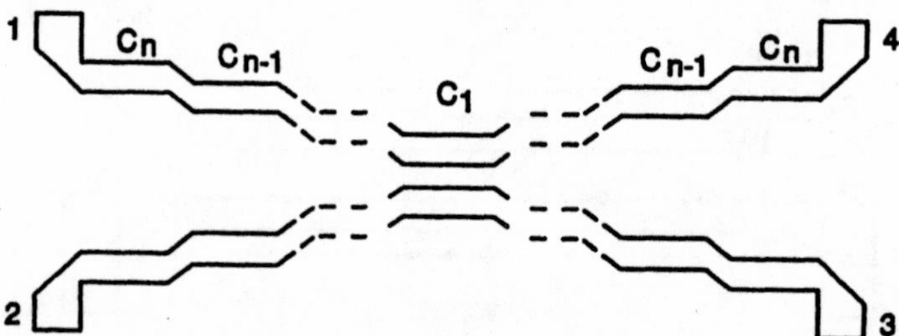

Figure 2.5. Uniform double-coupled line symmetrical multisection coupler.

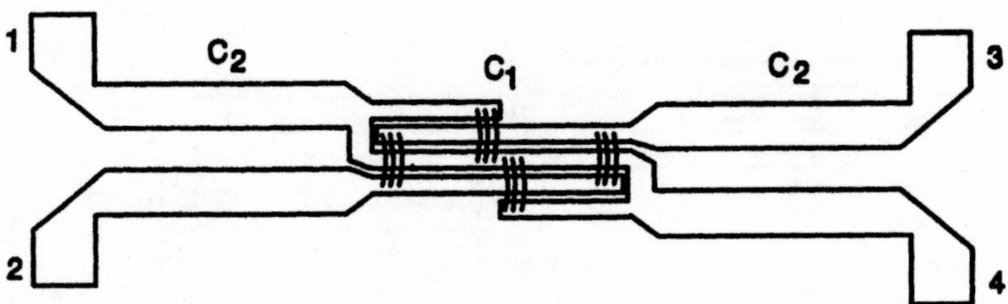

Figure 2.6. Uniform two- to four-line multisection coupler for tight coupling.

The next type of microstrip coupler, shown in Figure 2.6, can achieve wideband tight coupling, but is usually limited to three sections because of the prohibitively small dimensions required for the center section. The center section is a Lange coupler and may require some form of modification (a gap may be left between the ground plane and the substrate for this section) [22,23].

Tandem connection of two loosely double-coupled lines is an attractive choice for obtaining tight coupling [24]. Figure 2.7 is an illustration of this technique. The direct port of the first coupler is connected to the isolated port of the second coupler, and tandem connection is achieved by connecting the coupled port of the first coupler to the input port of the second coupler. Crossovers at the coupler centers may also be used to achieve easy tandem connection [21]. However, this requires a modification of the coupler center, which should be included in the analysis.

Theoretically it is not possible to achieve full-power (0 dB) coupling by using single, or multisection, or any other form of backward-wave coupler. However, this is possible by, again, the tandem connection of two -3 dB couplers. An ideal choice would be to employ two -3 dB Lange couplers [25] as illustrated in Figure 2.8.

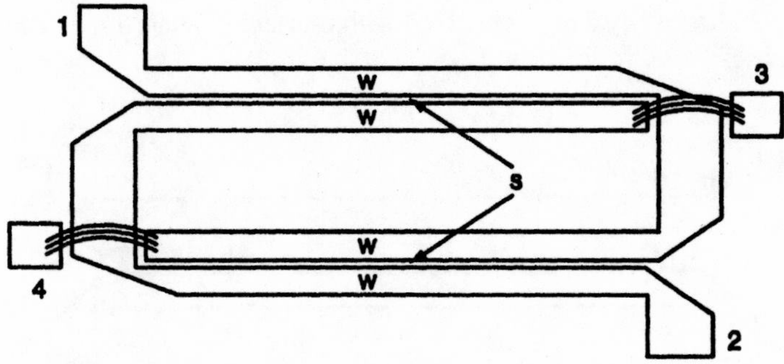

Figure 2.7. Uniform two-line tandem coupler.

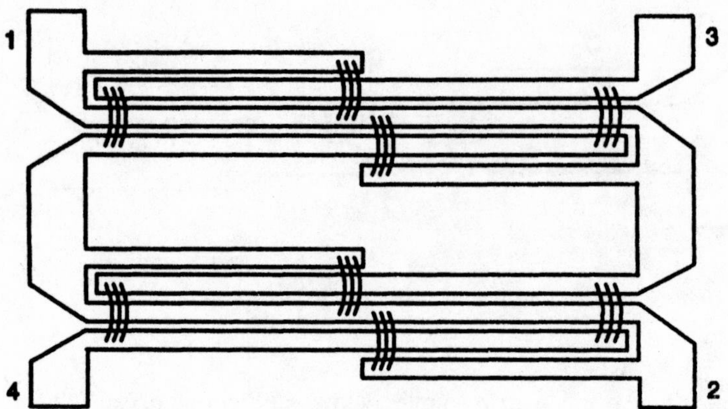

Figure 2.8. A 0-dB tandem Lange coupler.

Almost 0 dB (minus losses) occurs at the crossover points of the direct and coupled signals. Lange couplers are usually slightly overcoupled to increase the useable bandwidth. The overall isolation for a tandem coupler is less than the isolation of the individual couplers.

Until now we have discussed two basic types; namely, double-coupled and interdigitated Lange couplers. Lange couplers have acceptable directivities for many applications, but this is not true with double-coupled lines. The difference between the even- and odd-mode phase velocities increases with increasing frequency, which causes a sharp degradation of performance. Therefore, we must compensate for the difference between these velocities. This will be discussed next.

The simplest form of phase velocity equalization can be achieved by using input and output capacitances in the coupled region, as illustrated in Figure 2.9. The capacitances may be either lump chip capacitors or tightly coupled short fingers of microstrip lines. The latter may not give the required capacitance. Significant improvement in coupler directivity can be achieved with this technique [26,27]. The capacitors affect the phases of both direct and coupled signals.

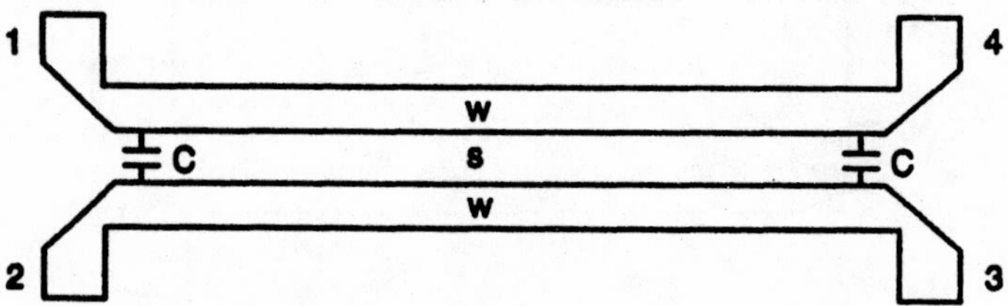

Figure 2.9. Capacitance-compensated uniform two-line coupler.

The most widely used phase velocity compensation technique employs a saw-tooth-type modification for the inner edges of coupled lines [28]. This wiggly coupler, as it is usually called, is shown in Figure 2.10. It is a very attractive form since phase velocity compensation is achieved by entirely planar means. The phase quadrature (which exists for all types of symmetrical couplers irrespective of the coupling level, provided there is a sufficiently good match and isolation) relationship between direct and coupled ports is not affected and is rather improved because directivity is improved.

An alternative to wiggly configuration, a serpentine coupler [29] as shown in Figure 2.11, can also be used for phase velocity compensation. The serpentine configuration can be derived from the wiggly coupler by rounding off the sharp edges.

Other forms of phase velocity compensation can be realized by overlay couplers [30], by suspended microstrip couplers, or by alternative designs, which will not be discussed. An alternative to the multisection coupler design is to employ continuously tapered coupled lines in which sharp transitions can be effectively eliminated. The basic shape of a nonuniform coupler is illustrated in Figure 2.12. Coupling coefficient increases from almost zero value from either end of the coupler, to its tightest value at the center of the coupler.

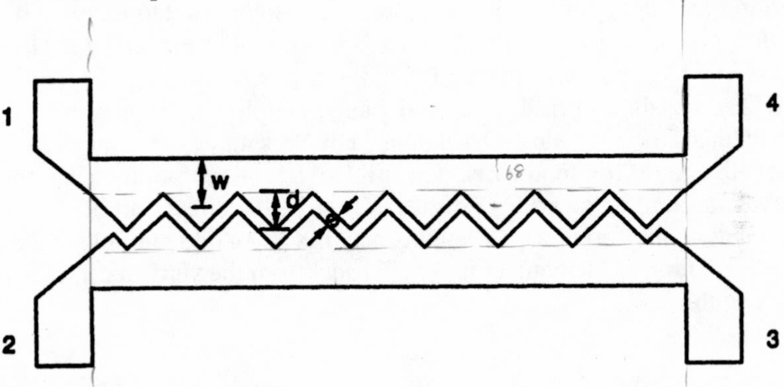

Figure 2.10. Wiggly two-line coupler.

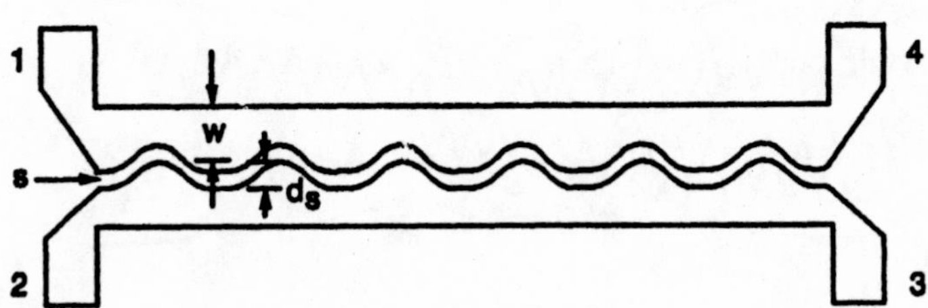

Figure 2.11. Serpentine two-line coupler.

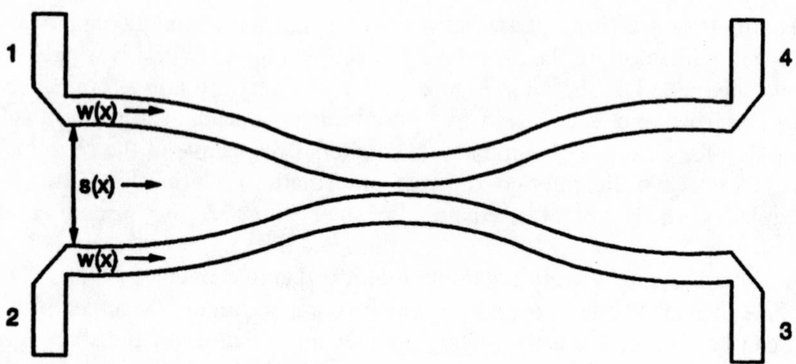

Figure 2.12. Nonuniform two-line coupler.

As it is, this coupler, like its multisection uniform version, does not provide a very useful bandwidth due to sharp deterioration in its directivity. The wiggly line-compensation technique for uniform couplers can also be used to improve directivity in nonuniform couplers [31]. Such a coupler is illustrated in Figure 2.13. This coupler can give very wideband performance for loose nominal coupling.

It is also possible to design tapered Lange couplers as shown in Figure 2.14. However, in this case, very long interdigitated couplers may not be practical due to the increased path lengths for bond wires. The original Lange or its unfolded versions may be employed, depending on which port is required as the direct port. It should also be noted here that nonuniform coupler lengths may not be an odd multiple of the quarter-wavelength. As long as the coupler is symmetrical from the vertical center axis it may have any length.

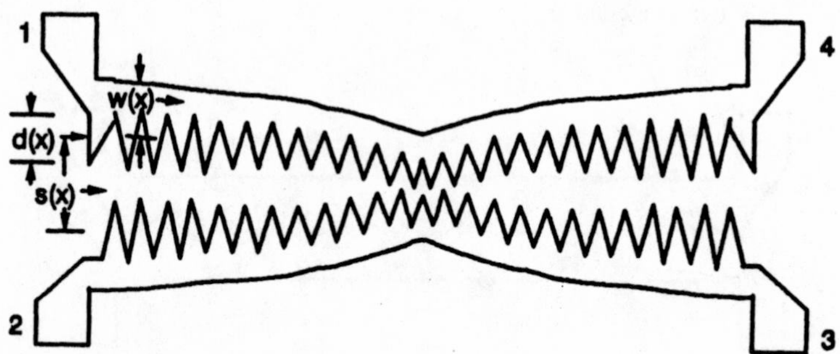

Figure 2.13. Nonuniform wiggly two-line coupler.

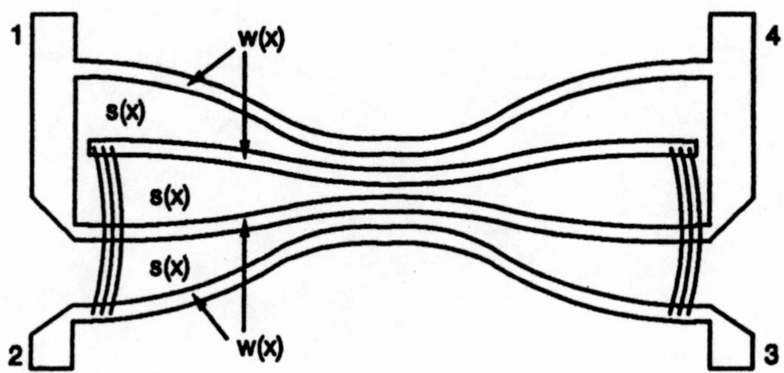

Figure 2.14. Nonuniform Lange coupler.

With moderate nominal multioctave coupling it should be sufficient to introduce a nonuniform Lange at the center of the coupler for the realization of relatively tight coupling coefficient values [31]. This form of coupler is shown in Figure 2.15.

The crossovers can be interchanged, depending on the application. Due to the transition from double-coupled lines to four lines, there is a sharp change in the velocity of propagation for both the even and odd modes. This must be considered during synthesis of the coupling function. Details will be given in Chapter 3.

For ultrawideband tight coupling we must connect two loosely coupled nonuniform couplers in tandem, as given in Figure 2.16. This configuration is an ideal choice for -3 dB ultrawideband backward-wave couplers using microstrip technology [31]. It can provide in excess of 10:1 -3 dB quadrature performance with acceptable directivity and insertion loss for many applications.

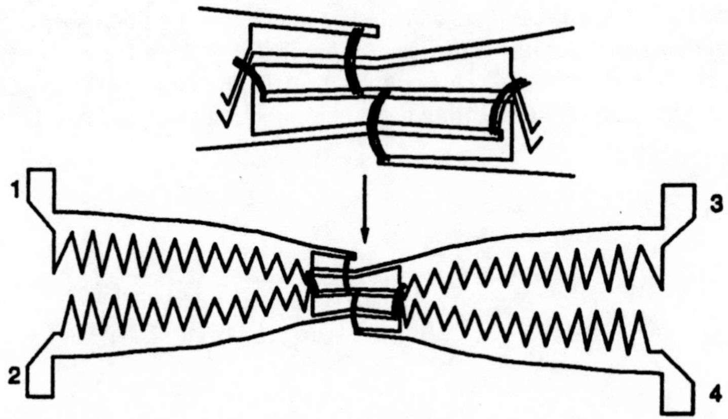

Figure 2.15. Nonuniform wiggly two- to four-line coupler.

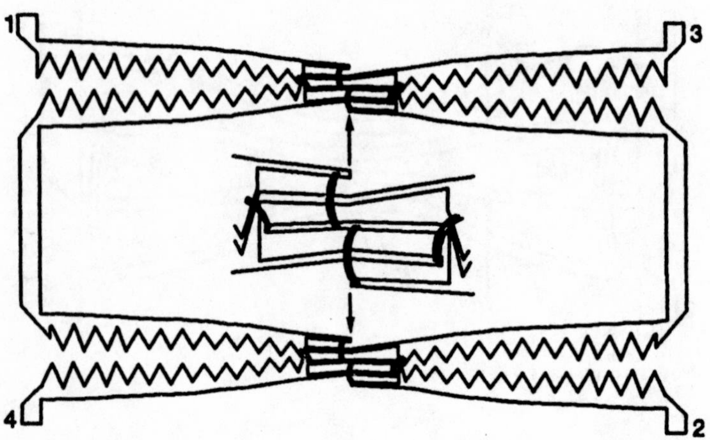

Figure 2.16. Nonuniform wiggly two- to four-line tandem coupler.

It was stated earlier that tight coupling on the order of -3 dB can be achieved by use of Lange or tandem couplers. However, these are not entirely planar because they require bond wires. With nonuniform coupled lines, it is possible to realize -3 dB or tighter coupling values with moderate bandwidths by using entirely planar (no bond wires) configurations with easily realizable gaps [32]. This novel coupler is illustrated in Figure 2.17. It requires a suitable modification of the reflection coefficient distribution function associated with nonuniform lines, which will be discussed in detail in Chapter 6. We can see from Figure 2.17 that such a modification results in a rather distinctive wavy shape for the nonuniform coupled lines.

Last, but not least, forward-wave (also called *codirectional* because the direct and coupled ports are adjacent) couplers can also be realized using microstrip technology [33–35]. Bandpass coupler principles can be applied to design forward-wave couplers with very high directivities. This coupler utilizes the phase velocity difference, and similar modifications like wiggling or serpentining may be employed (in this case

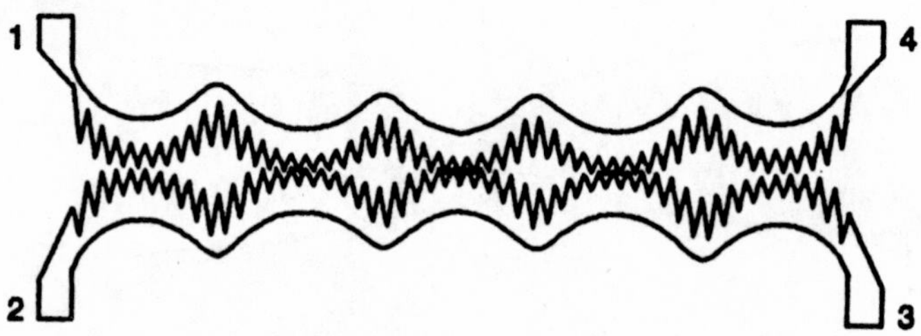

Figure 2.17. Bandpass-type nonuniform two-line coupler.

to increase the phase velocity difference) [36]. The simplest configuration for such a coupler is given in Figure 2.18. The coupling function for this coupler is a function of the coupler length, which may be taken long enough to realize 0 dB coupling.

Table 2.1 summarizes the microstrip directional couplers discussed so far.

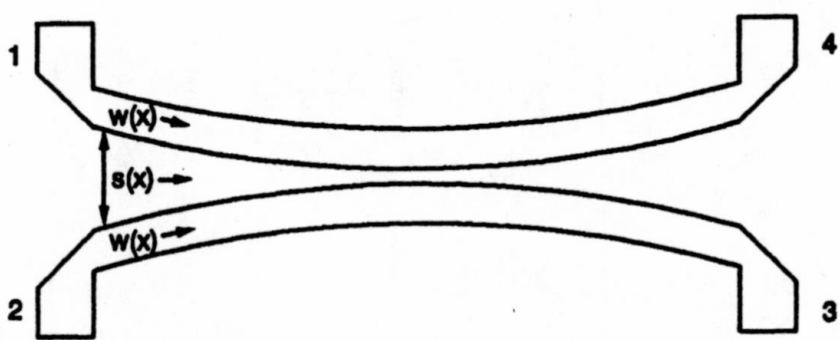

Figure 2.18. Nonuniform codirectional coupler.

2.5 TYPES OF MICROSTRIP FILTERS

There are basically two types of microstrip filters: transmission line filters utilizing short or open circuited capacitive or inductive stubs and filters utilizing coupling principles of edge- or end-coupled lines [37–46]. This is a rather coarse generalization because many MIC circuits, whether intentional or otherwise, often behave like a filter. Filtering requirements may significantly vary for different applications. Therefore, it is not recommended to draw specific guidelines or prepare universal specification sheets for microstrip filters. However, as a rule of thumb, it is generally required that filter circuits introduce very little or no attenuation for the desired signal. It is also desirable to have sharp skirts and high out-of-band rejection. Some common filter circuits, together with the ones discussed in Chapters 7 and 8, are shown in Figure 2.19.

2.6 APPLICATIONS OF DIRECTIONAL COUPLERS AND FILTERS

Microstrip directional couplers have been used extensively in microwave networks and systems, satellite and mobile communications, and antennas and radar systems. They form the basis of high-quality performance circuits because of their matched characteristics at all four ports. They perform a wide variety of functions, such as combining and splitting power in transmitters, receivers, modulators, and amplifiers; sampling power for level control; matching input and output of balanced amplifiers; providing dc isolation; and passive phase shifting in a number of applications. Some of these applications and others will be analyzed in detail in Chapter 10.

Table 2.1
Symmetrical microstrip directional couplers

Type	*Coupling* (dB)*	*Bandwidth*	*Directivity*	*Manufacturing Difficulty*	*Coupling Sensitivity*
Uniform two-line Figure 2.2	−7	Narrow	Poor	Very easy	Low
Uniform three-line Figure 2.3	−4	Moderate	Fair	Moderate	Low
Uniform Lange Figure 2.4	−2.5	Octave	Good	Moderate	Moderate
Uniform multisection two-line Figure 2.5	−10	—	Poor	Easy	Low
Uniform multisection two- to four-line Figure 2.6	−8	Moderate	Fair	Moderate	Moderate
Tandem two-line Figure 2.7	−3	Narrowband	Poor	Easy	Low
Tandem Lange Figure 2.8	0	Octave	Fair	Moderate	Moderate
Capacitance compensated two-line Figure 2.9	−7	Narrow	Very good	Easy	Moderate
Wiggly two-line Figure 2.10	−7	Narrow	Very good	Very easy	Low

Serpentine two-line Figure 2.11	−7	Narrow	Good	Easy	Low
Nonuniform two-line Figure 2.12	−10	—	Poor	Easy	Moderate
Nonuniform wiggly two-line Figure 2.13	−10	Wide	Good	Easy	Moderate
Nonuniform Lange Figure 2.14	−8	Wide	Fair	Moderate	High
Nonuniform two- to four-line Figure 2.15	−8	Ultrawide	Good	Moderate	High
Nonuniform two- to four-line tandem Figure 2.16	−3	Ultrawide	Good	Moderate	High
Bandpass two-line Figure 2.17	−0.5	Moderate	Good	Easy	Moderate
Codirectional Smooth edge Figure 2.18	0	Narrow	Excellent	Easy	—

*Approximate tightest coupling on alumina substrate.

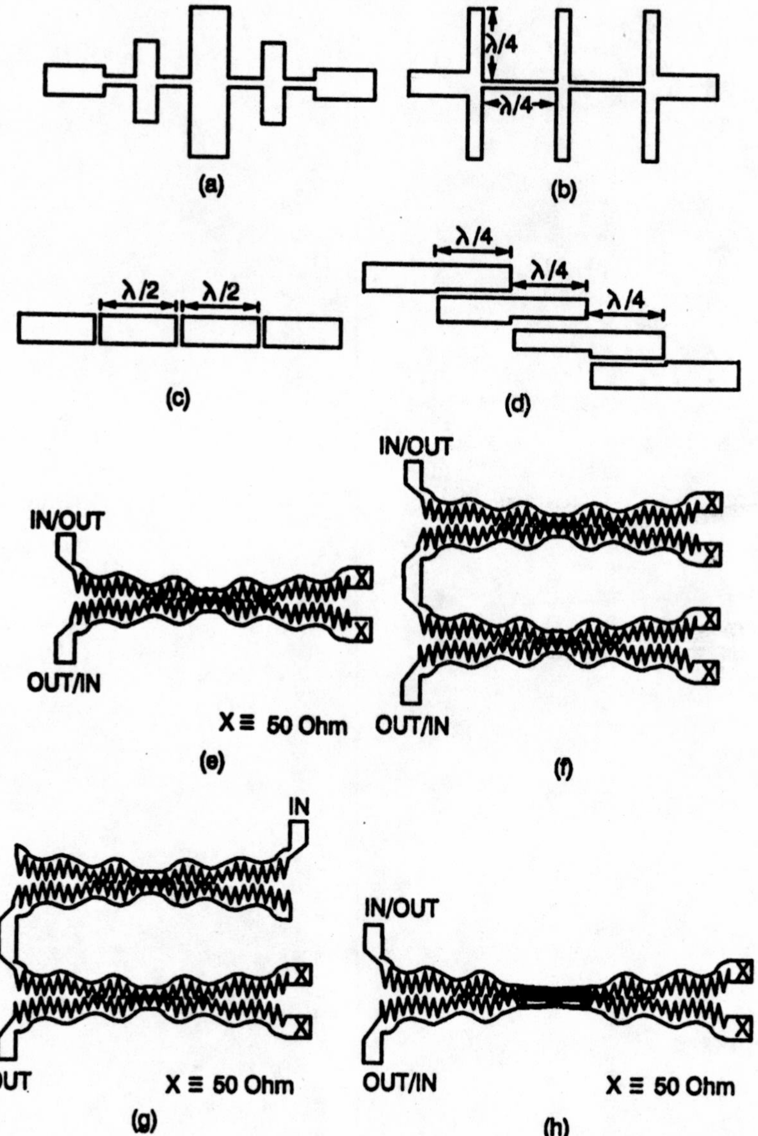

Figure 2.19. Microstrip transmission line and coupled-line filters: (a) low-pass, (b) bandstop, (c) end-coupled bandpass, (d) edge-coupled bandpass, (e) matched nonuniform coupled-line bandpass, (f) matched nonuniform coupled-line high-rejection bandpass, (g) nonuniform coupled-line high-rejection bandpass, and (h) nonuniform coupled-line periodic.

Whenever there is a need to pass a certain band of signals and to reject the rest, a filter is used. Among their several functions are channel cleaning and multiplexing an input signal. Although they are lossy, passive microstrip filters are easy to design, can be very compact, have less volume, are light weight, and have highly repeatable performance. They are also widely used at the receiver front ends, for IF extraction in mixers, and in modems.

REFERENCES

[1] Pucel, R. A., D. J. Masse, and C. P. Hartwig, "Losses in Microstrip," *IEEE Trans. on Microwave Theory and Tech.,* Vol. MTT-16, 1968, pp. 342–350.

[2] Rao, B. R., "Effect of Loss and Frequency Dispersion on the Performance of Microstrip Directional Couplers and Coupled Line Filters," *IEEE Trans. on Microwave Theory and Tech.,* Vol. MTT-22, 1974, pp. 747–750.

[3] Lewin, L., "Spurious Radiation from Microstrip," *Proc. IEE,* Vol. 125, 1978, pp. 633–642.

[4] Majewski, M. L., R. W. Rose, and J. R. Scott, "Modeling and Characterization of Microstrip-to-Coaxial Transitions," *IEEE Trans. on Microwave Theory and Tech.,* Vol. MTT-29, 1981, pp. 799–807.

[5] Green, H. E., "The Numerical Solution of Some Important Transmission-Line Problems," *IEEE Trans. on Microwave Theory and Tech.,* Vol. MTT-13, 1965, pp. 676–692.

[6] Smith, J. I., "The Even- and Odd-Mode Capacitance Parameters for Coupled Lines in Suspended Substrate," *IEEE Trans. on Microwave Theory and Tech.,* Vol. MTT-19, 1971, pp. 424–431.

[7] Bryant, T. G., and I. A. Weiss, "Parameters of Microstrip Transmission Lines and of Coupled Pairs of Microstrip Lines," *IEEE Trans. on Microwave Theory and Tech.,* Vol. MTT-16, 1968, pp. 1021–1027.

[8] Kirschning, M., and R. H. Jansen, "Accurate Wide-Range Design Equations for the Frequency-Dependent Characteristic of Parallel Coupled Microstrip Lines," *IEEE Trans. on Microwave Theory and Tech.,* Vol. MTT – 32, 1984, pp. 83–90.

[9] Hammerstad, E., and Ø. Jensen, "Accurate Models for Microstrip Computer-Aided Design," *IEEE MTT-S Int. Microwave Symp. Digest,* 1980, pp. 407–409.

[10] Akhtarzad, S., T. R. Rowbotham, and P. B. Johns, "The Design of Coupled Microstrip Lines," *IEEE Trans. on Microwave Theory and Tech.,* Vol. MTT-23, 1975, pp. 486–492.

[11] Napoli, L. S., and J. J. Hughes, "Characteristics of Coupled Microstrip Lines," *RCA Review,* September 1970, pp. 479–498.

[12] Edwards, T. C., *Foundations for Microstrip Circuit Design,* 2nd Ed., New York: John Wiley & Sons, 1991.

[13] Perlow, S. M., and A. Presser, "The Interdigitated Three-Strip Coupler," *IEEE Trans. on Microwave Theory and Tech.,* Vol. MTT – 32, 1984, pp. 1418–1422.

[14] March, S. L., "Simple Equations Characterize Bond Wires," *Microwaves & RF,* November 1991, pp. 105–110.

[15] Lange, J., "Interdigitated Stripline Quadrature Hybrid," *IEEE Trans. on Microwave Theory and Tech.,* Vol. MTT-17, 1969, pp. 1150–1151.

[16] Presser, A., "Interdigitated Microstrip Coupler Design," *IEEE Trans. on Microwave Theory and Tech.,* Vol. MTT-26, 1978, pp. 801–805.

[17] Waugh, R., and D. LaCombe, "Unfolding the Lange Coupler," *IEEE Trans. on Microwave Theory and Tech.,* Vol. MTT-20, 1972, pp. 777–779.

[18] Rizzoli, V., and A. Lipparini, "The Design of Interdigitated Couplers for MIC Applications," *IEEE Trans. on Microwave Theory and Tech.,* Vol. MTT-26, 1978, pp. 7–15.

[19] Kajfez, D., Z. Paunovic, and S. Pavlin, "Simplified Design of Lange Coupler," *IEEE Trans. on Microwave Theory and Tech.,* Vol. MTT-26, 1978, pp. 806–808.

[20] Paolino, D. D., "Design More Accurate Interdigitated Couplers," *Microwaves,* May 1976, pp. 34–38.

[21] Miley, J. E., "Looking For a 3 to 8 dB Microstrip Coupler," *Microwaves,* March 1974, pp. 58–62.

[22] Tajima, Y., and S. Kamihashi, "Multiconductor Couplers," *IEEE Trans. on Microwave Theory and Tech.,* Vol. MTT-26, 1978, pp. 795–801.

[23] Tserng, H. Q., and S. R. Nelson, "5-18 GHz, 3 dB Hybrid Couplers," *Electronics Lett.,* Vol. 17, 1981, pp. 258–259.

[24] Shelton, J. P., J. Wolfe, and R. C. V. Wagoner, "Tandem Couplers and Phase Shifters for Multi-Octave Bandwidth," *Microwaves,* April 1965, pp. 14–19.

[25] Waugh, R. W., "Sensitivity Analysis of the Lange Coupler," *Microwave J.,* November 1989, pp. 121–129.

[26] Kajfez, D., "Raise Coupler Directivity With Lumped Compensation," *Microwaves,* March 1978, pp. 64–70.

[27] March, S. L., "Phase Velocity Compensation in Parallel-Coupled Microstrip," *IEEE MTT-S Int. Microwave Symp. Digest,* 1982, pp. 410–412.

[28] Podell, A., "A High Directivity Microstrip Coupler Technique," *IEEE G-MTT Symp. Digest,* Vol. 33, 1970, pp. 33–36.

[29] De Ronde, F. C. "Wide-Band High Directivity in MIC Proximity Couplers by Planar Means," *IEEE MTT-S Int. Microwave Symp. Digest,* 1984, pp. 480–482.

[30] Paolino, D. D., "MIC Overlay Coupler Design Using Spectral Domain Techniques," *IEEE Trans. on Microwave Theory and Tech.,* Vol. MTT-26, 1978, pp. 646–649.

[31] Uysal, S., and A. H. Aghvami, "Synthesis, Design and Construction of Ultra-Wide-Band Nonuniform Quadrature Directional Couplers in Inhomogeneous Media," *IEEE Trans. on Microwave Theory and Tech.,* Vol. MTT – 37, 1989, pp. 969–976.

[32] Uysal, S., and J. Watkins, "Novel Microstrip Multifunction Directional Couplers and Filters for Microwave and Millimeter-Wave Applications," *IEEE Trans. on Microwave Theory and Tech.,* Vol. MTT – 39, 1991, pp. 977–985.

[33] Ikalainen, P. K., and G. L. Matthaei, "Wide-Band, Forward-Coupling Microstrip Hybrids with High Directivity," *IEEE Trans. on Microwave Theory and Tech.,* Vol. MTT – 35, 1987, pp. 719–725.

[34] Gunton, D. J., "Design of Wideband Co-directional Couplers and Their Realization at Microwave Frequencies Using Coupled Comblines," *Microwaves, Optics and Acoustics,* Vol. 2, 1978, pp. 19–30.

[35] Islam, S., "A New Analytic Design Technique for Two- and Three-way Warped Mode Combline Directional Couplers," *IEEE Trans. on Microwave Theory and Tech.,* Vol. MTT – 37, 1989, pp. 34–42.

[36] Uysal, S., and J. Watkins, "Forward-Wave Nonuniform Microstrip Couplers," *Proc. of 21st European Microwave Conf.,* Vol. 1, 1991, pp. 722–727.

[37] Matthaei, G. L., L. Young, and E. M. T. Jones, *Microwave Filters, Impedance Matching Networks, and Coupling Structures,.* New York: McGraw-Hill, 1964.

[38] Malherbe, J. A. G., *Microwave Transmission Line Filters,* Dedham, MA: Artech House, 1979.

[39] Kirton, P. A., and K. K. Pang, "Extending the Realizable Bandwidth of Edge-Coupled Stripline Filters," *IEEE Trans. on Microwave Theory and Tech.,* Vol. MTT-25, 1977, pp. 672–676.

[40] Cristal, E. G., "Tapped-Line Coupled Transmission Lines with Applications to Interdigital and Combline Filters," *IEEE Trans. on Microwave Theory and Tech.,* Vol. MTT-23, 1975, pp. 1007–1012.

[41] Cohn, S. B., "Parallel Coupled Transmission-line Resonator Filters," *IRE Trans. Microwave Theory and Tech.,* Vol. 6, 1958, pp. 223–231.

[42] Bui, L. Q., Y. C. Shih, and T. N. Ton, "mm-Wave Harmonic-Reject Filter," *Microwave J.,* July 1987, pp. 119–122.

[43] Cristal, E. G., and S. Frankel, "Hairpin-Like and Hybrid Line/Half-Wave Parallel-Coupled-Line Filters," *IEEE Trans. on Microwave Theory and Tech.,* Vol. MTT-20, 1972, pp. 719–728.

[44] Moazzam, M. R., S. Uysal, and A. H. Aghvami, "Improved Performance Parallel Coupled Filters," *Microwave J.,* Vol. 11, 1991, pp. 127–135.

[45] Jokela, K. T., "Narrow-Band Stripline or Microstrip Filters with Transmission Zeros at Real and Imaginary Frequencies," *IEEE Trans. on Microwave Theory and Tech.,* Vol. MTT-28, 1980, pp. 542–547.

[46] Walker, J. L. B., "Exact and Approximate Synthesis of TEM-Mode Transmission-Type Directional Filters," *IEEE Trans. on Microwave Theory and Tech.,* Vol. MTT-26, 1978, pp. 186–192.

Chapter 3
Synthesis of Nonuniform Line Directional Couplers

3.1 INTRODUCTION

The design of nonuniform line directional couplers requires a relationship between the geometry of the coupled line and the continuous coupling coefficient function ($k(x)$). At present a direct synthesis of the continuous physical dimensions is not available. Instead, the uniform coupled-lines data are assumed to be valid at each elemental section of the nonuniform coupled lines [1]. The transformation from uniform to nonuniform data is achieved by the use of inverse cubic spline interpolation. For a detailed formulation of cubic spline functions the reader may refer to [2]. Figure 3.1 gives an illustration of the transformation procedure.

We assume that the nonuniform coupler is made up of infinitesimally small lengths of uniform couplers as shown in Figure 3.1(b). The synthesized coupling coefficient function for the nonuniform coupler may resemble to the one given in Figure 3.1(c). The initial gap s_1 usually corresponds to almost no coupling; that is, $k(-l/2) = 0$. This restriction will be removed in Chapter 6. Because we have uniform coupler parameters as functions of the coupling coefficient, which are defined as cubic spline functions, Figure 3.1(d), we can then evaluate these functions at every value of $k(x)$ to obtain the continuous parameters for the nonuniform coupler. The functions g_1, g_2, g_3 represent cubic spline functions. This shows that, instead of forcing a single polynomial to pass through all the data points, we have $N - 1$ cubic functions for N data points. Accuracy of this transformation can be maintained within 0.001% of any parameter with only about 30 data points specified for the uniform coupler parameter. This technique also allows us to go back and forth from uniform to nonuniform data.

3.2 SYNTHESIS OF UNIFORM COUPLED-LINE PARAMETERS

In Section 1.4, we derived an equation for the optimization of shape ratios for a given substrate and impedance level. This equation is

$$Z_0^2 c^2 C_{ea} C_{oa} \sqrt{\epsilon_{re}(f)\epsilon_{ro}(f)} - 1 = 0 \tag{3.1}$$

Where Z_0 is the characteristic impedance (usually 50Ω), $c = 300 \times 10^9$ mm/s, C_{ea} and C_{oa} are even- and odd-mode capacitances, respectively, with dielectric replaced by air, and $\epsilon_{re}(f)$ and $\epsilon_{ro}(f)$ are the frequency-dependent effective dielectric constants for the respective modes. The input parameters required for the optimization are substrate thickness h, relative dielectric constant ϵ_r, characteristic impedance Z_0, spacing between the conductors s, and design frequency f.

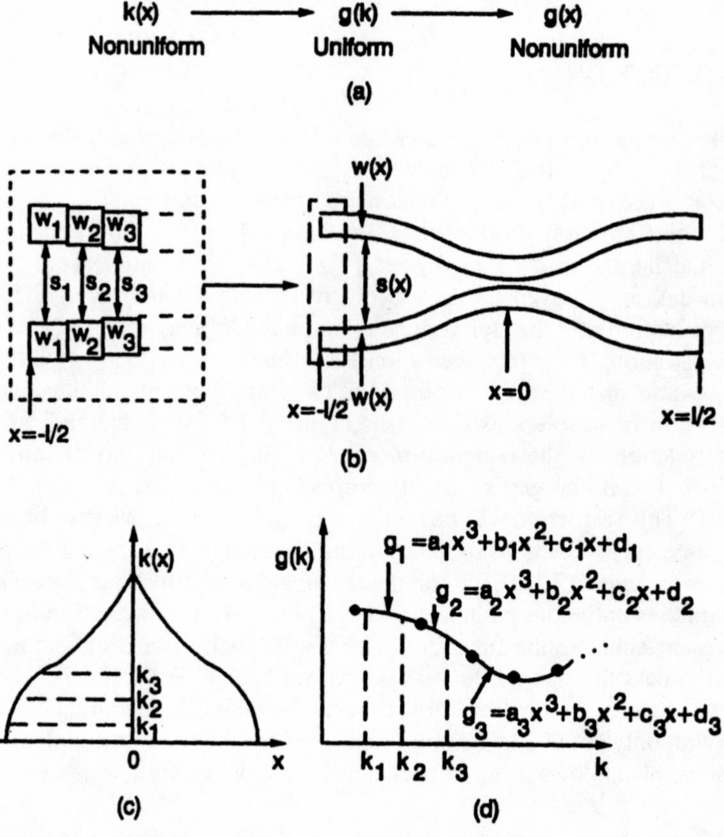

Figure 3.1. Transformation from uniform data to nonuniform data: (a) transformation form continuous-coupling coefficient to a physical parameter, (b) transformation of elemental values to continuous curves, (c) typical continuous-coupling coefficient, and (d) inverse cubic spline interpolation.

The even- and odd-mode characteristic impedances and phase velocities are then given by

$$Z_{0e} = \frac{\sqrt{\epsilon_{re}(f)}}{cC_e} \qquad Z_{0o} = \frac{\sqrt{\epsilon_{ro}(f)}}{cC_o} \tag{3.2}$$

$$v_e = \frac{c}{\sqrt{\epsilon_{re}(f)}} \qquad v_o = \frac{c}{\sqrt{\epsilon_{ro}(f)}} \tag{3.3}$$

The coupling coefficient can be expressed in terms of the characteristic impedances for the two modes:

$$k = \frac{Z_{0e} - Z_{0o}}{Z_{0e} + Z_{0o} - \dfrac{2Z_{0e}Z_{0o}(N-2)}{(N-1)(Z_{0e} + Z_{0o})}} \tag{3.4}$$

where N is the number of conductors.

3.3 PHASE VELOCITY COMPENSATION

We have seen in Chapter 1 that a significant portion of the odd-mode field is between the conductors, both in air and dielectric. On the other hand, the even-mode field is confined to the dielectric. This asymmetrical division of fields will give rise to different mode velocities, which causes a degradation in directivity of the directional couplers. The odd-mode phase velocity can be slowed down to be equal to the even-mode phase velocity by providing extra distance for the odd-mode wave to travel. This can be done by wiggling the inner edges of the coupled lines as suggested by Podell [3]. Podell used a cut-and-try (intuitive) technique to arrive at an optimum wiggle depth for a given coupling. This was based on the fact that the effect of wiggling on the even-mode capacitance parameter C_{fe} is negligible for the tighter couplings achievable with double-doupled lines. This fact can be observed in Figure 1.5(a) and (b). The ratio C_{fo}/C_{fe} is greater than 7 for $k = 0.3$ and greater than 20 for $k = 0.4$. Therefore, for coupling values greater than 0.3 we can neglect the effect of wiggling on the even-mode phase velocity.

In the case of nonuniform coupled lines, coupling coefficient extends from very small values ($C_{fe} \approx C_{fo}$) at the coupler ends (assuming symmetrical couplers) to very high values ($C_{fe} \ll C_{fo}$) at the coupler center. This means that especially with ultrawideband designs in which a significant portion of the coupled lines are weakly coupled, the effect of wiggling on the even mode becomes significant and should be included in the design.

A semiempirical technique [4] has been reported for the determination of wiggle depth for phase velocity compensation. Consider the wiggly coupled lines as illustrated in Figure 3.2 The desired wiggling is introduced in such a way that the parallel plate capacitance C_p is not affected. We can also deduce from Figure 1.5(a) that even a significant change in conductor width ($k > 0.5$) has very little effect on the fringing capacitance C_f. Therefore, the odd-mode capacitance with and without wiggling can be expressed as

$$C_{ow} = C_p + C_f + C'_{fo} \tag{3.5}$$

$$C_o = C_p + C_f + C_{fo} \tag{3.6}$$

where the capacitances have their usual meaning, with C'_{fo} being the odd-mode capacitance between the conductors with wiggle.

We require the effective dielectric constants with wiggle to be equal; therefore,

$$\epsilon_{rew} = \epsilon_{row} \tag{3.7}$$

Neglecting, for the time being, the effect of wiggling on the even mode, we have

$$\epsilon_{re} = \epsilon_{row} \tag{3.8}$$

Therefore, it is sufficient to increase the odd-mode capacitance by a factor $\epsilon_{re}/\epsilon_{ro}$. Multiplying (3.6) by this factor and equating it to (3.5), we can solve for C'_{fo}:

$$C'_{fo} = C_{pf}\left(\frac{\epsilon_{re}}{\epsilon_{ro}} - 1\right) + \frac{\epsilon_{re}}{\epsilon_{ro}} C_{fo} \tag{3.9}$$

where $C_{pf} = C_p + C_f$.

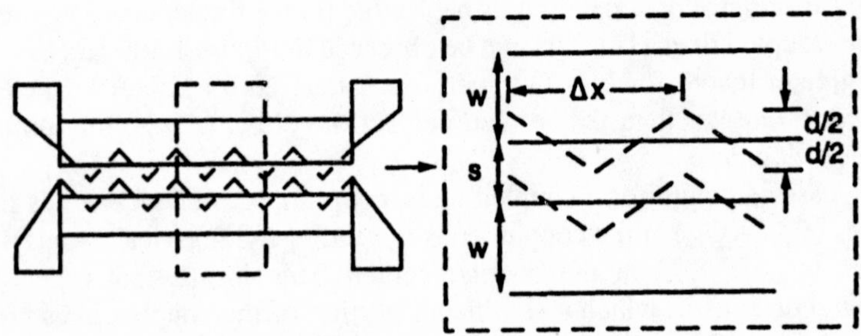

Figure 3.2. Introduction of wiggle into the coupled region.

The odd-mode length will be increased by the factor C'_{fo}/C_{fo}. From Figure 3.2 we can then obtain the required wiggle depth d:

$$d = \frac{\Delta x}{2}\sqrt{\left(\frac{C'_{fo}}{C_{fo}}\right)^2 - 1} \tag{3.10}$$

We find that, after rigorous computations and comparison with experimental data, an empirical modification to (3.10) is sufficient to include the effect of wiggling on the even mode:

$$d = \frac{\Delta x}{2}\exp[1.38C_{fe}/C_{fo}]\sqrt{\left(\frac{C'_{fo}}{C_{fo}}\right)^2 - 1} \tag{3.11}$$

This equation gives very accurate results for a wide range of coupling coefficient. The tested accuracy was found to be better than 2% for $k \geq 0.02$ on alumina substrate with $\epsilon_r = 9.9$.

The compensated phase velocity is then modified as

$$v = \frac{v_e}{\exp(0.06C_{fe}/C_{fo})} \tag{3.12}$$

3.4 COMPUTED DESIGN CURVES FOR SOME IMPORTANT SUBSTRATES

In this section, design curves in the form of cubic splines will be given for some practical common substrates including the emerging superconductive ones. Because surface waves [5] become a major problem at higher frequencies, we have divided design curves according to a given substrate thickness. In all cases we shall use $Z_0 = 50\ \Omega$. Design curves will include $w(k)$ (conductor width versus coupling coefficient), $s(k)$ (conductor spacing versus coupling coefficient), $d(k)$ (wiggle depth versus coupling coefficient), and $v(k)$ (compensated phase velocity versus coupling coefficient). In some cases, instead of $v(k)$ we shall give design curves for $v_e(k)$ and $v_o(k)$ (e.g., for interdigitated couplers).

3.4.1 Design Curves for Alumina Substrate with $\epsilon_r = 9.9$ and $h = 0.635$ mm

The design curves shown in Figure 3.3 are computed from dc to 20 GHz in steps of 5 GHz. Intermediate values can be obtained by interpolation. The dispersive nature of coupled lines can easily be seen from Figure 3.3(a). At $k = 0.02$, the strip width w

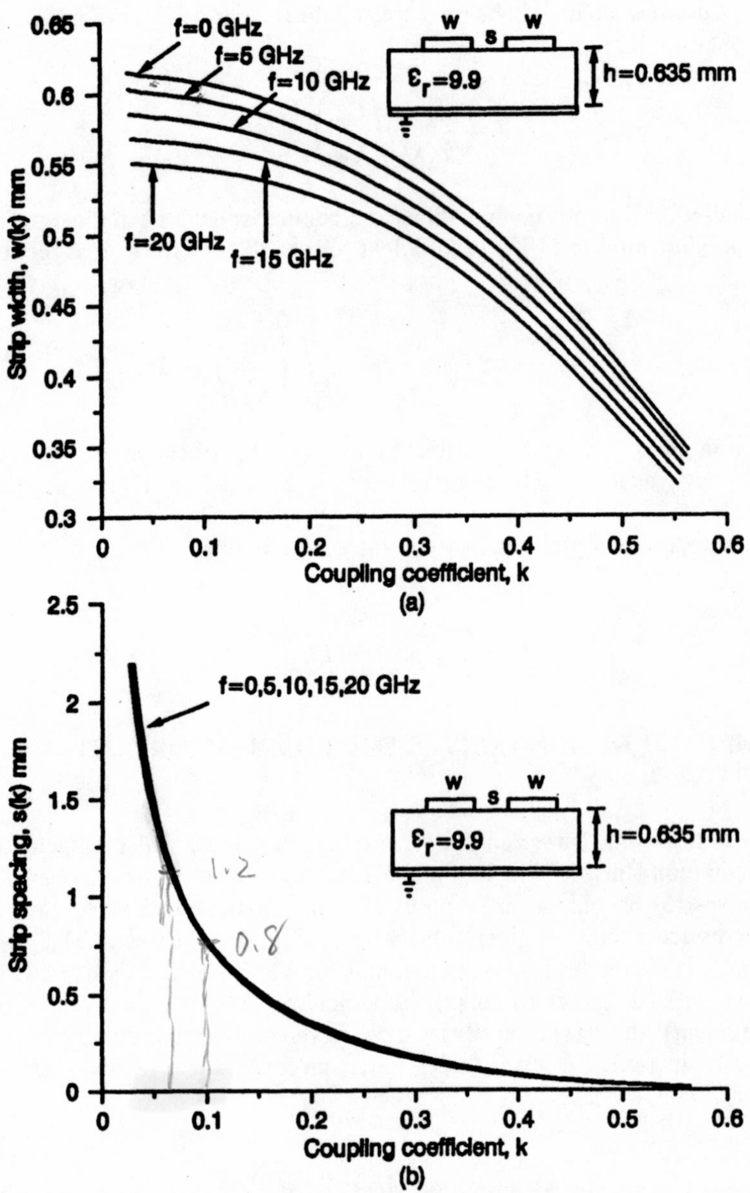

Figure 3.3. Design curves for double-coupled lines on alumina substrate with $\epsilon_r = 9.9$, $h = 0.635$ mm, and $f = 0\text{–}20$ GHz: (a) strip width $w(k)$ and (b) strip spacing $s(k)$.

decreases from 0.619 mm to 0.557 mm when frequency is increased from dc to 20 GHz. Design curves for the gap $s(k)$ are shown in Figure 3.3(b). Because the variation in s is very small, the curves overlap. It is simply not practical to plot $s(k)$ to show its variation with frequency; we would need about 10 plots.

The required wiggle depth for phase velocity compensation is given in Figure 3.4(a). For tighter coupling ($k > 0.12$), the wiggle depth increases with increasing frequency. Because at $k = 0$ we have $v_e = v_o$, no wiggle is required; that is, $d = 0$. This fact is confirmed in the computations, and there is observed a point of inflection at about $k = 0.06$. The selection of Δx depends on the width of the conductors. In practice we use $\Delta x = 0.3$ mm for 0.635-mm thick alumina substrate. As d is directly proportional to Δx (see equation (3.11)), any other value corresponding to a different Δx can be obtained simply by multiplying the values given in Figure 3.4(a) by the desired factor.

The compensated phase velocity is given in Figure 3.4(b). The reduction in phase velocity is higher for weak coupling because wiggling affects both the even and odd modes. Variation in phase velocity with frequency is almost constant for all coupling values. With no wiggling the even-mode phase velocity at $k = 0.06$ and $f = 0$ is 114×10^9 mm/s; with wiggling this value is around 107×10^9 mm/s.

3.4.2 Design Curves for Alumina Substrate with $\epsilon_r = 9.9$ and $h = 0.25$ mm

The effect of surface waves with 0.25-mm thick substrate can be neglected up to around 60 GHz. The next set of design curves are computed for $f = 25, 35, 45,$ and 55 GHz. Owing to its smaller thickness, this substrate is less dispersive. The strip width versus coupling coefficient is given in Figure 3.5(a). The maximum variation in strip width is about 0.01 mm from 25 to 55 GHz. The design gap for this substrate is plotted for the same frequency and coupling range and is shown in Figure 3.5(b). The design curves for $d(k)$ and $v(k)$ are given in Figure 3.6.

3.4.3 Design Curves for Alumina Substrate with $\epsilon_r = 9.9$ and $h = 0.1$ mm

The next set of design curves are particularly suitable for directional couplers and filters at 77 and 95 GHz millimeterwave radar subsystems. The design curves for $w(k)$ and $s(k)$ are given in Figure 3.7. We can see that dispersion in negligibly small. Figure 3.8(a) gives the required wiggle depth for phase velocity compensation. In this case, it is very difficult to maintain the required dimensional accuracy with ordinary photolithographic techniques. However, it is possible to realize phase velocity compensated coupled lines with e-beam deposition (writing) technique on 0.1-mm thick alumina substrate up to 100 GHz. Because the required fractional bandwidth is much lower

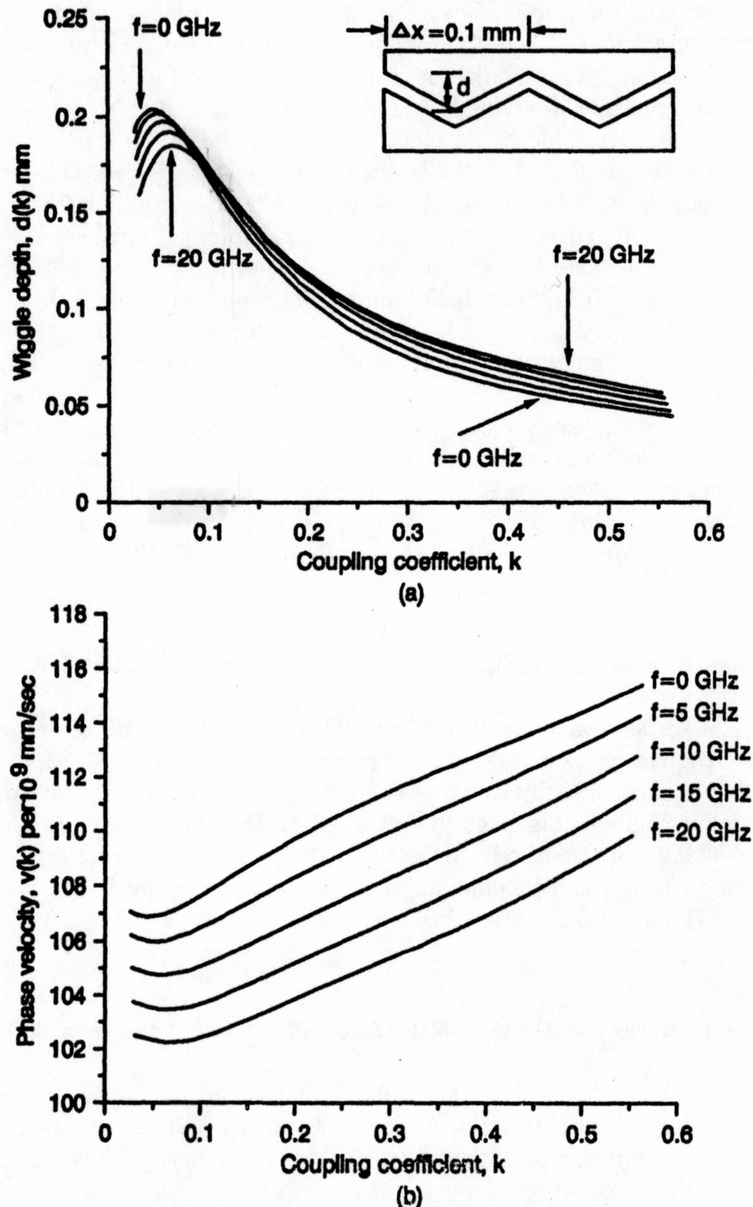

Figure 3.4. Design curves for double-coupled lines on alumina substrate with $\epsilon_r = 9.9$, $h = 0.635$ mm, and $f = 0$–20 GHz: (a) wiggle depth $d(k)$ and (b) compensated phase velocity $\upsilon(k)$.

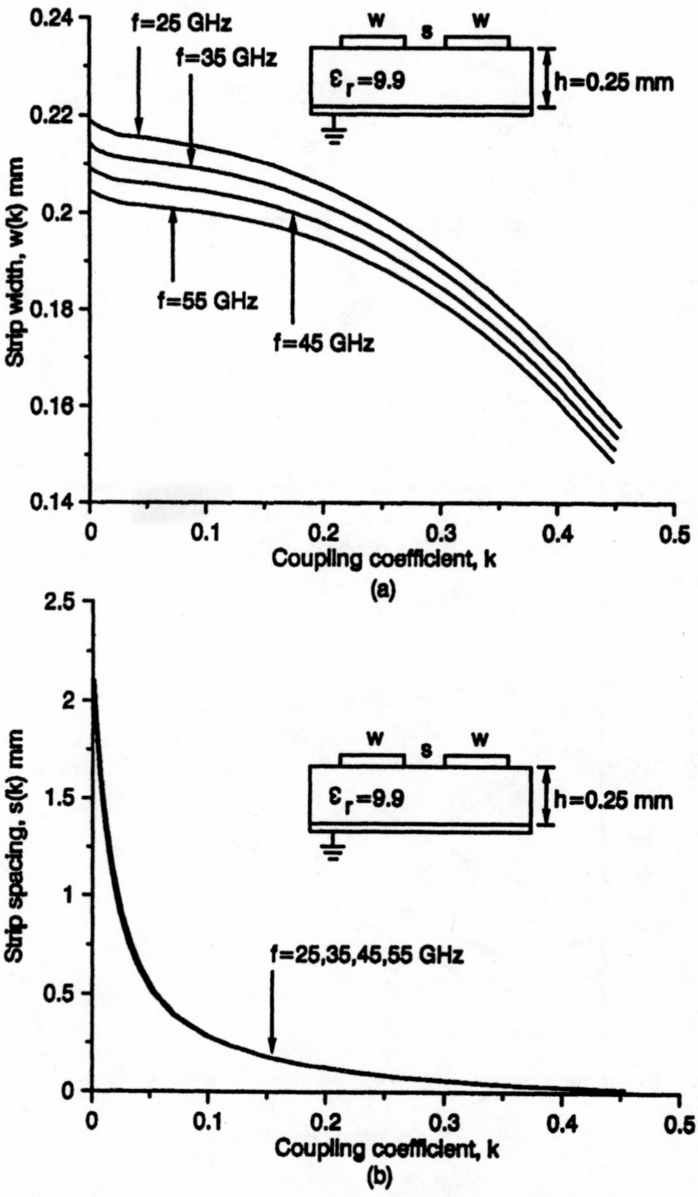

Figure 3.5. Design curves for double-coupled lines on alumina substrate with $\epsilon_r = 9.9$, $h = 0.25$ mm, and $f = 25–55$ GHz: (a) strip width $w(k)$ and (b) strip spacing $s(k)$.

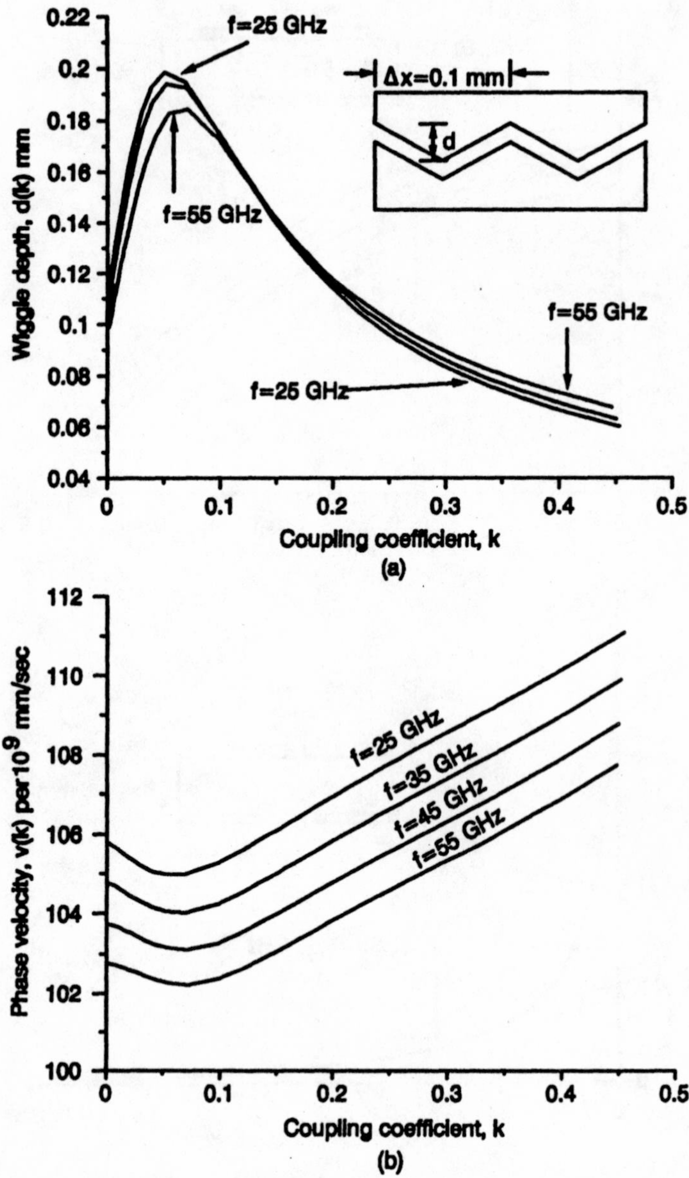

Figure 3.6. Design curves for double-coupled lines on alumina substrate with $\epsilon_r = 9.9$, $h = 0.25$ mm, and $f = 25$–55 GHz: (a) wiggle depth $d(k)$ and (b) compensated phase velocity $v(k)$.

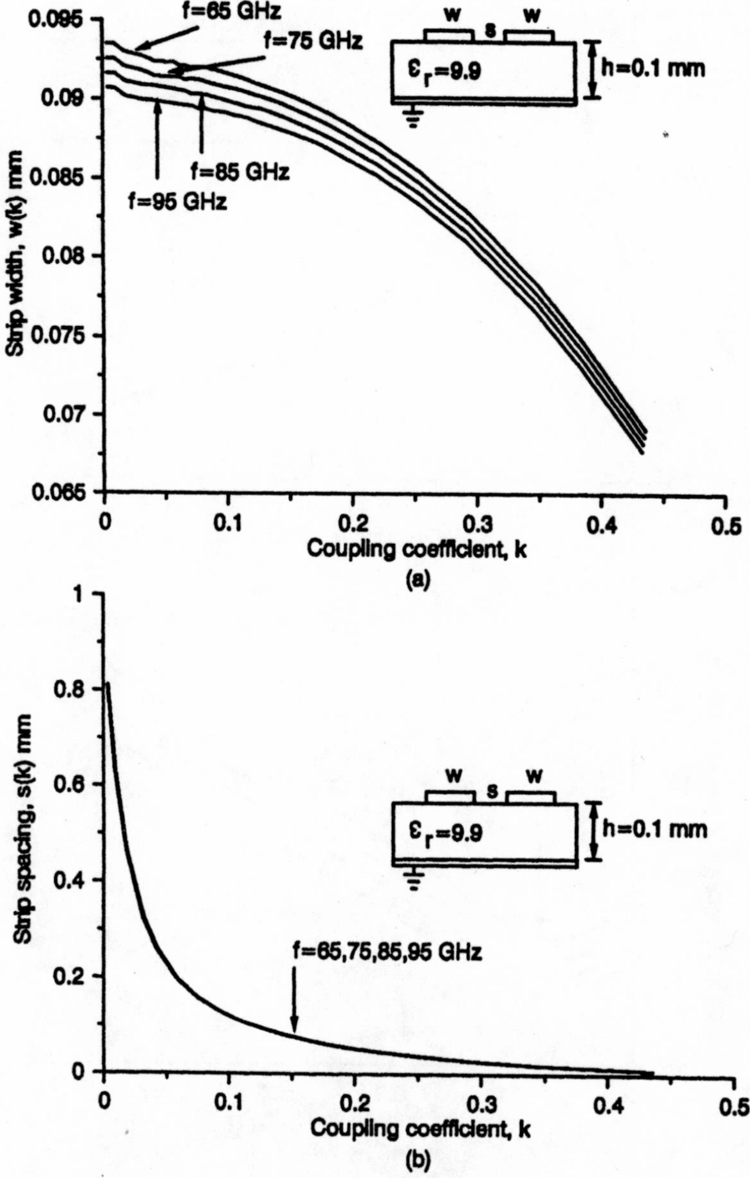

Figure 3.7. Design curves for double-coupled lines on alumina substrate with $\epsilon_r = 9.9$, $h = 0.1$ mm, and $f = 65$–95 GHz: (a) strip width $w(k)$ and (b) strip spacing $s(k)$.

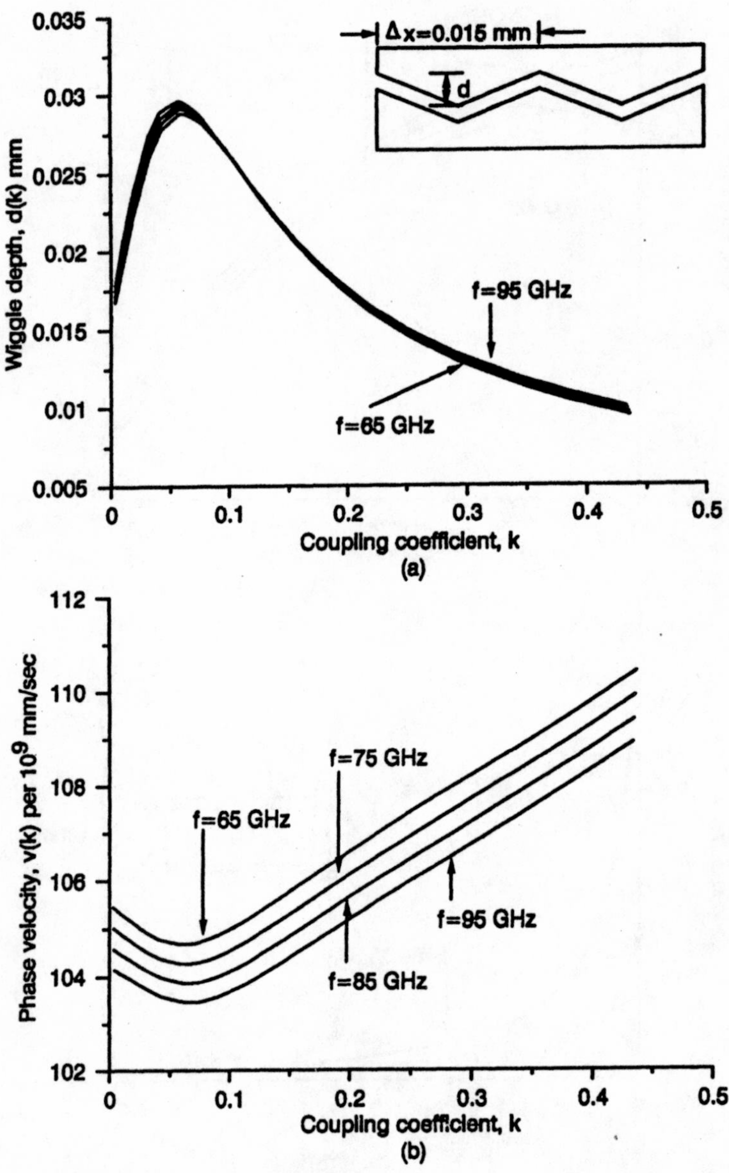

Figure 3.8. Design curves for double-coupled lines on alumina substrate with $\epsilon_r = 9.9$, $h = 0.1$ mm, and $f = 65–95$ GHz: (a) wiggle depth $d(k)$ and (b) compensated phase velocity $v(k)$.

compared to microwave frequencies, the use of codirectional couplers may be a better alternative.

In this section the design curves are computed at $f = 65, 75, 85,$ and 95 GHz.

3.4.4 Design Curves for Interdigitated Couplers on Alumina Substrate with $\epsilon_r = 9.9$ and $h = 0.635$ mm

In this section, we present design curves for interdigitated (four-conductor) couplers. Figure 3.9 gives design curves for $w(k)$ and $s(k)$. Both of these couplers are usually employed for tight coupling (e.g., -3 dB). However, in the design of ultrawideband nonuniform directional couplers, an interdigitated section is required at the coupler center for coupling coefficient values not realizable by double-coupled lines; that is, from -7 to -2.5 dB ($k = 0.45$ to $k = 0.75$). The use of an interdigitated center section reduces the number of loosely coupled tandem couplers from three or more to two.

For interdigitated couplers, phase velocity compensation becomes extremely complex and will not be considered. Therefore, in this case, we give the even- and odd-mode phase velocities shown in Figure 3.10 as part of the design. We will need these velocities when designing ultrawideband directional couplers. This will be discussed in Section 3.5.

3.4.5 Design Curves for GaAs Substrate with $\epsilon_r = 12.9$ and $h = 0.2$ mm

Monolithic microwave integrated circuit technology offers large-scale integration of microwave components on a single chip. This increasingly important technology also offers circuits with smaller chip sizes and better performance because in many cases bond wires are modeled as underpasses or air bridges.

Design curves for GaAs substrate with $h = 0.2$ mm, cover 5–35 GHz. Strip width is plotted as a function of the coupling coefficient in steps of 10 GHz, from 5 to 35 GHz, and is given in Figure 3.11(a). Maximum dispersion in w for this range is about 10 μm. Figure 3.11(b) gives the function $s(k)$; in this case the effect of frequency is negligibly small.

The required wiggle depth for phase velocity compensation is computed with $\Delta x = 0.03$ mm and is given in Figure 3.12 (a). Because d is directly proportional to Δx, different values for wiggle depth can be computed directly from the simple geometry shown in the inset diagram. Using GaAs substrate provides about 12% length reduction for directional couplers compared to alumina substrate with the same thickness.

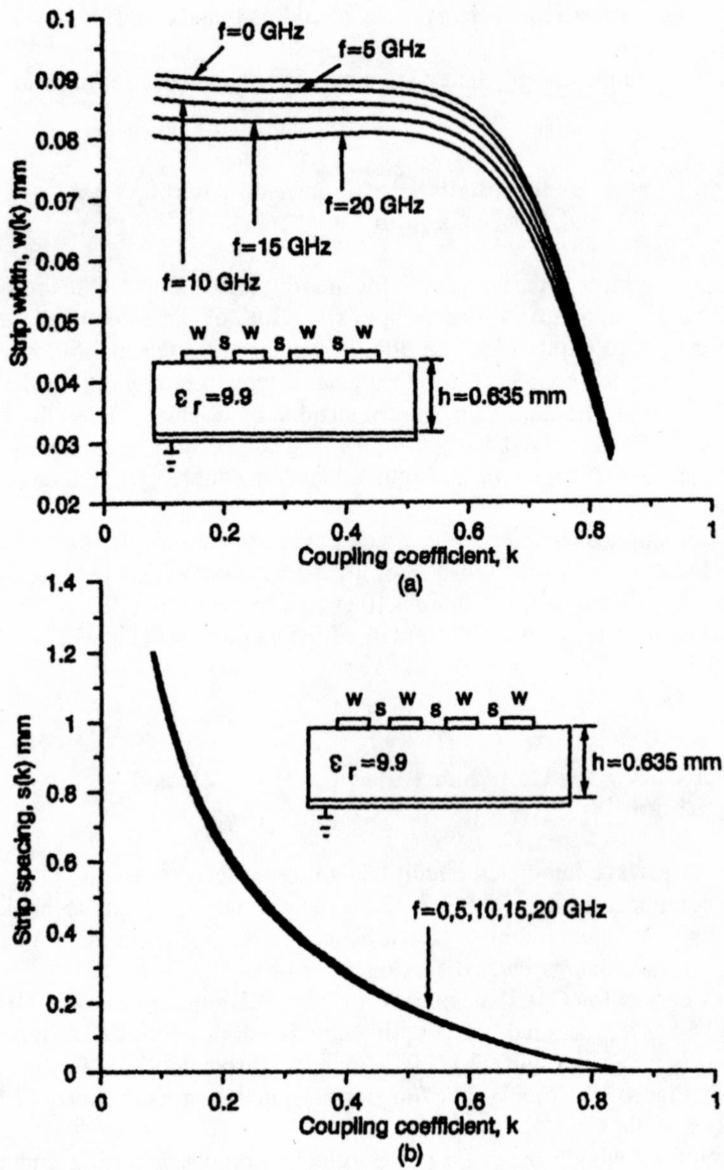

Figure 3.9. Design curves for interdigitated couplers on alumina substrate with $\epsilon_r = 9.9$, $h = 0.635$ mm, and $f = 0$–20 GHz: (a) strip width $w(k)$ and (b) strip spacing $s(k)$.

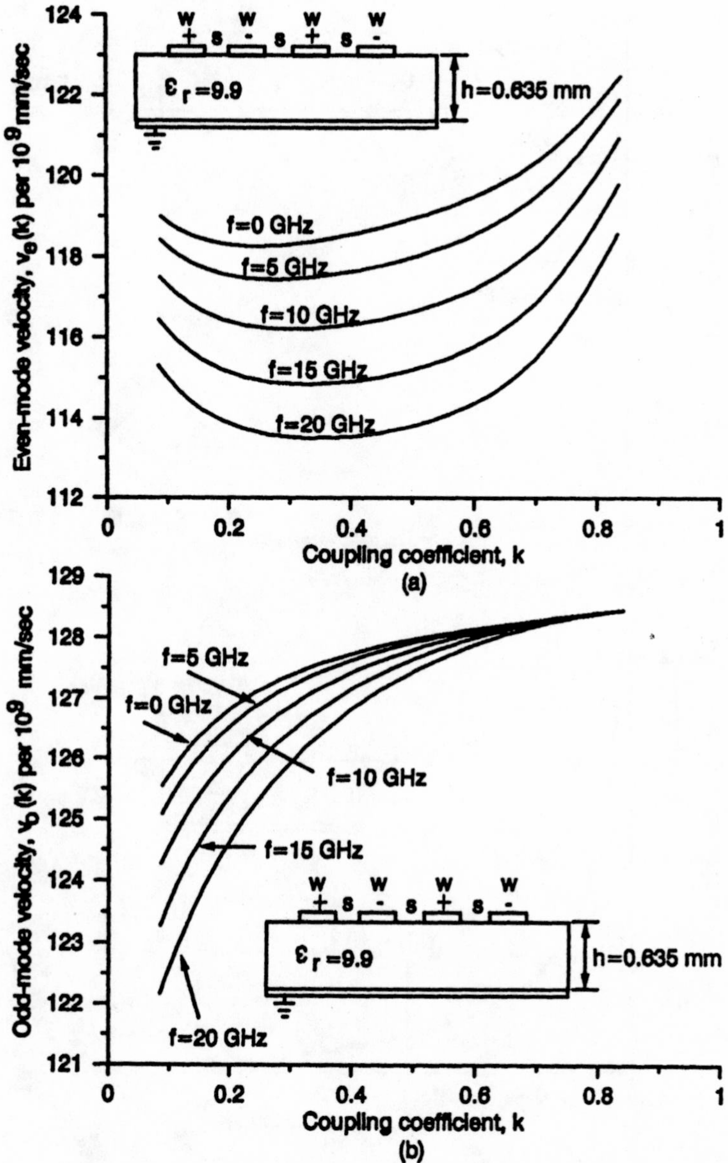

Figure 3.10. Design curves for interdigitated couplers on alumina substrate with $\epsilon_r = 9.9$, $h = 0.635$ mm, and $f = 0$–20 GHz: (a) even-mode phase velocity $v_e(k)$ and (b) odd-mode phase velocity $v_o(k)$.

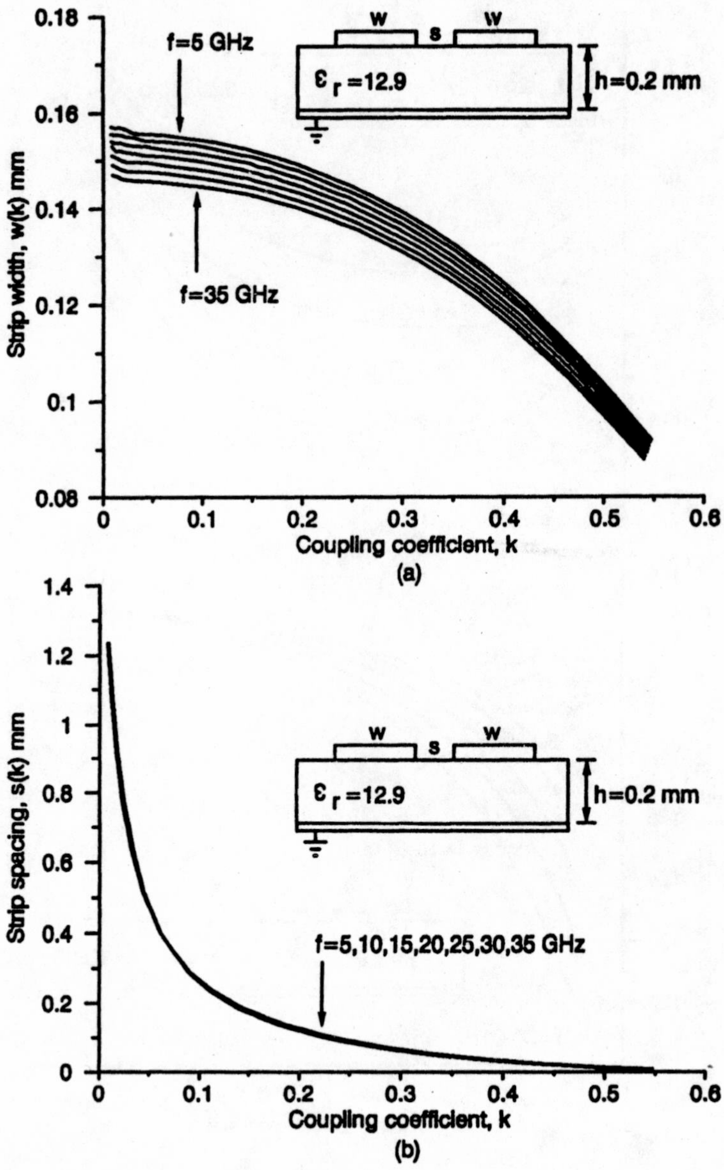

Figure 3.11. Design curves for double-coupled lines on GaAs substrate with $\epsilon_r = 12.9$, $h = 0.2$ mm, and $f = 5$–35 GHz: (a) strip width $w(k)$ and (b) strip spacing $s(k)$.

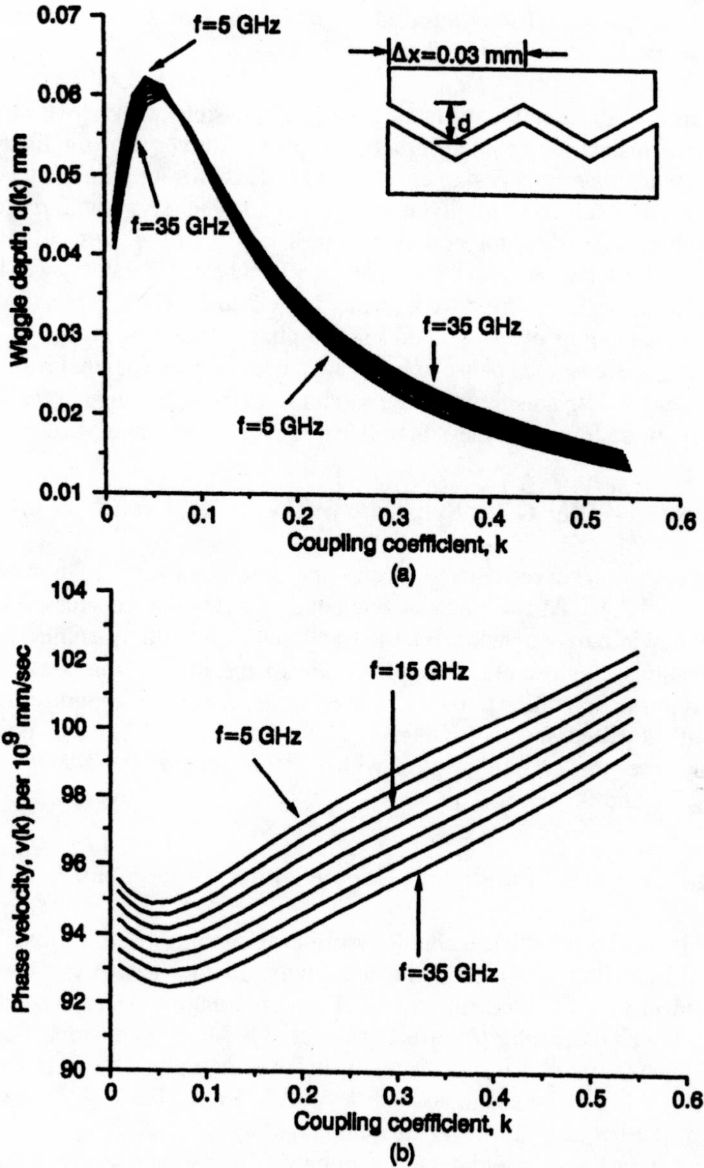

Figure 3.12. Design curves for double-coupled lines on GaAs substrate with $\epsilon_r = 12.9$, $h = 0.2$ mm, and $f = 5\text{–}35$ GHz: (a) wiggle depth $d(k)$ and (b) compensated phase velocity $v(k)$.

3.4.6 Design Curves for Interdigitated Couplers on GaAs Substrate with $\epsilon_r = 12.9$ and $h = 0.2$ mm

Very tight control of manufacturing tolerances is possible with MMIC technology. This allows the realization of interdigitated couplers with very narrow linewidths and gaps on GaAs substrate. The design curves $w(k)$ and $s(k)$ for interdigitated couplers on 0.2-mm thick GaAs substrate are given in Figure 3.13. There is about 2 μm oscillation in the computed linewidths for coupling values less than 0.4. For coupling values greater than 0.4 the tested accuracy (compared to Touchstone™ values) was better than 0.35%. The design curves cover frequencies from 5 to 55 GHz in steps of 10 GHz.

The corresponding even- and odd-mode phase velocities are given in Figure 3.14. Variation in even-mode phase velocity with coupling coefficient for the range 0.4 to 0.7 is less than 1.4%. This figure is the same for a 20 GHz change in frequency. For the odd mode, these figures are less than 0.3% for the same range of coupling values.

3.4.7 Design Curves for GaAs Substrate with $\epsilon_r = 12.9$ and $h = 0.1$ mm

The next set of design curves cover a much wider frequency range. These are given in Figures 3.15 and 3.16. At millimeter wavelengths, as we have discussed before, the frequency bandwidth may be wide, but the fractional bandwidth is small. Therefore, in many applications single-section couplers can do the job. In this section we give compensated phase velocities that can be used in the design of nonuniform bandpass filters, as will be discussed in Chapter 7. However, the new class of codirectional couplers can give any coupling value with suitable lengths for MMIC realization especially at 77 and 95 GHz.

3.4.8 Design Curves for Lanthalum Gallate with $\epsilon_r = 24.5$ and $h = 0.25$ mm

The emerging superconductive technology offers minimum insertion loss for microstrip coupled-line directional couplers and filters. In this case we have made an exception and used a shielded microstrip. The computations were carried out with $b = 2.5$ mm, which has negligible effect on the coupled-line dimensions. Ganguly and Krowne [6] have reported a full wave analysis covering up to 100 GHz for a pair of coupled lines on Lanthalum Gallate with $h = 0.254$ mm. For a given spacing and frequency we predict slightly lower coupling.

Design curves for strip width and spacing using this substrate are given in Figure 3.17. Even- and odd-mode phase velocities are given in Figure 3.18. It is a known fact that strong magnetic fields caused by sharp features (wiggles) may destroy superconductivity. Therefore, at this stage we do not consider phase velocity compensation in superconductive coupled lines. However, it should be possible to employ serpentine couplers in which sharp wiggles can effectively be eliminated. Alternatively, nonuni-

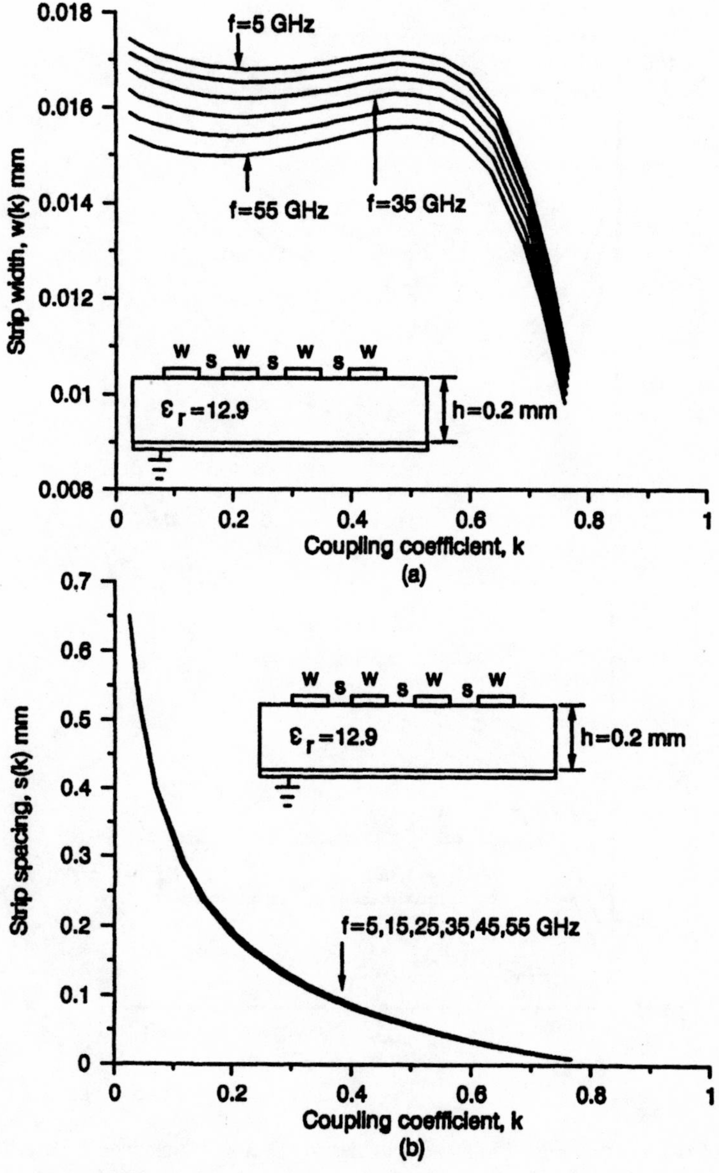

Figure 3.13. Design curves for interdigitated couplers on GaAs substrate with $\epsilon_r = 12.9$, $h = 0.2$ mm, and $f = 5$–55 GHz: (a) strip width $w(k)$ and (b) strip spacing $s(k)$.

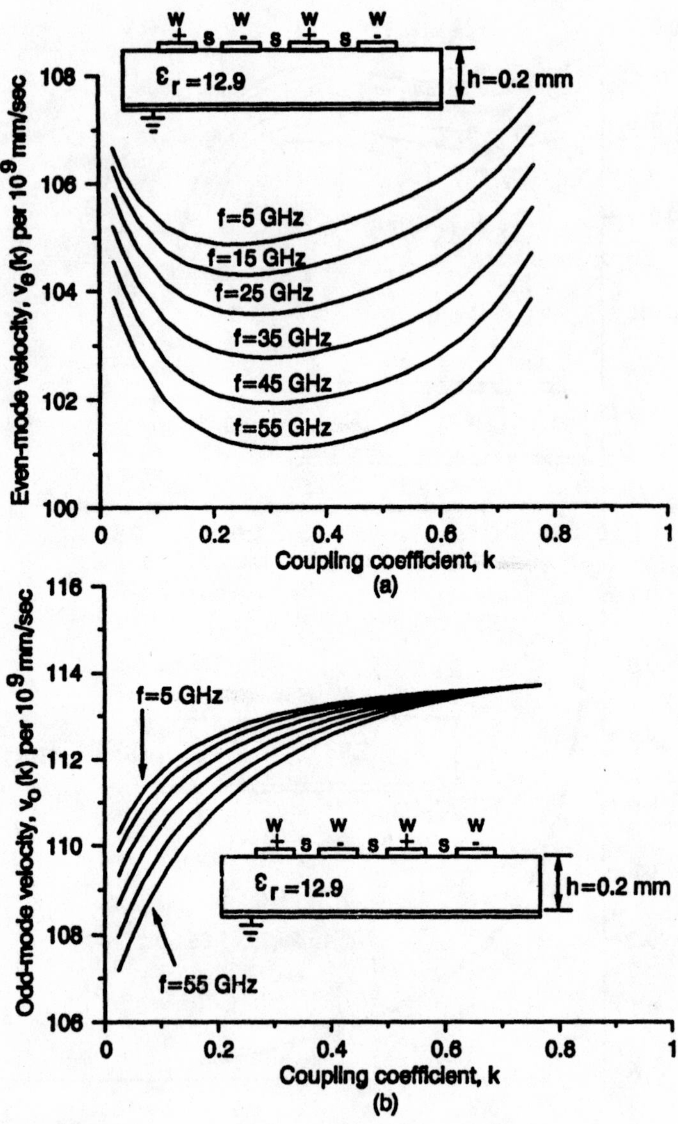

Figure 3.14. Design curves for interdigitated couplers on GaAs substrate with $\epsilon_r = 12.9$, $h = 0.2$ mm, and $f = 5–55$ GHz: (a) even-mode phase velocity $v_e(k)$ and (b) odd-mode phase velocity $v_o(k)$.

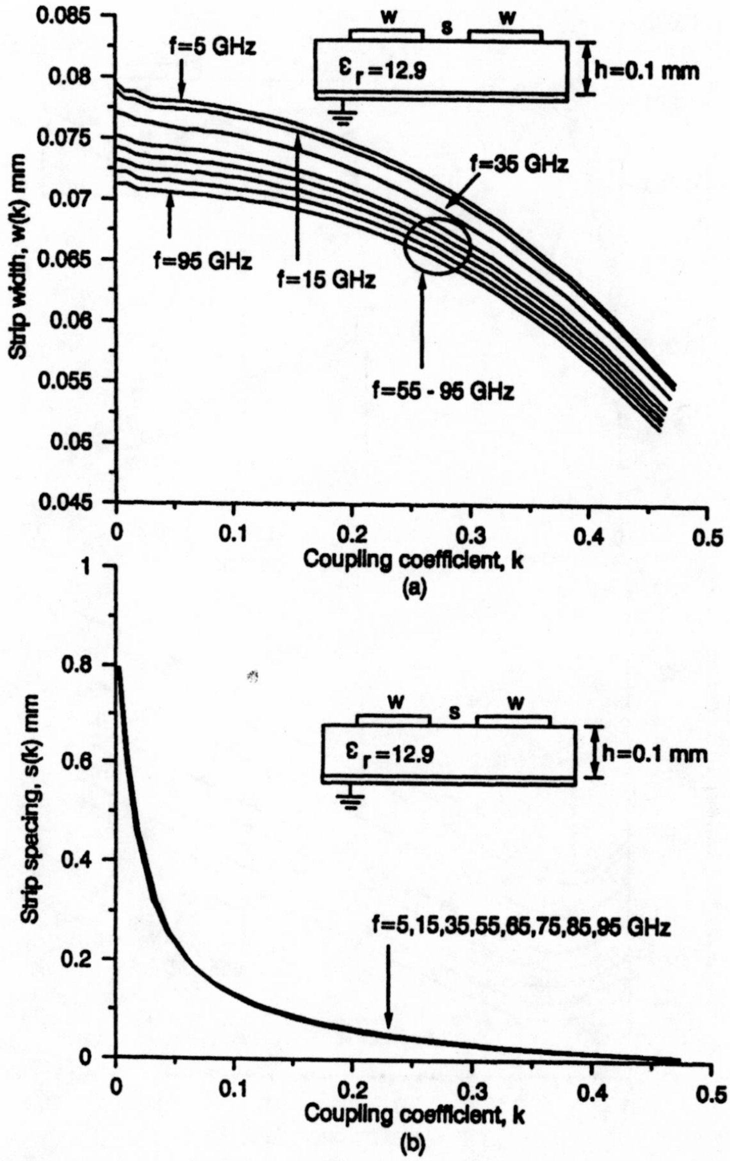

Figure 3.15. Design curves for double-coupled lines on GaAs substrate with $\epsilon_r = 12.9$, $h = 0.1$ mm, and $f = 5$–95 GHz: (a) strip width $w(k)$ and (b) strip spacing $s(k)$.

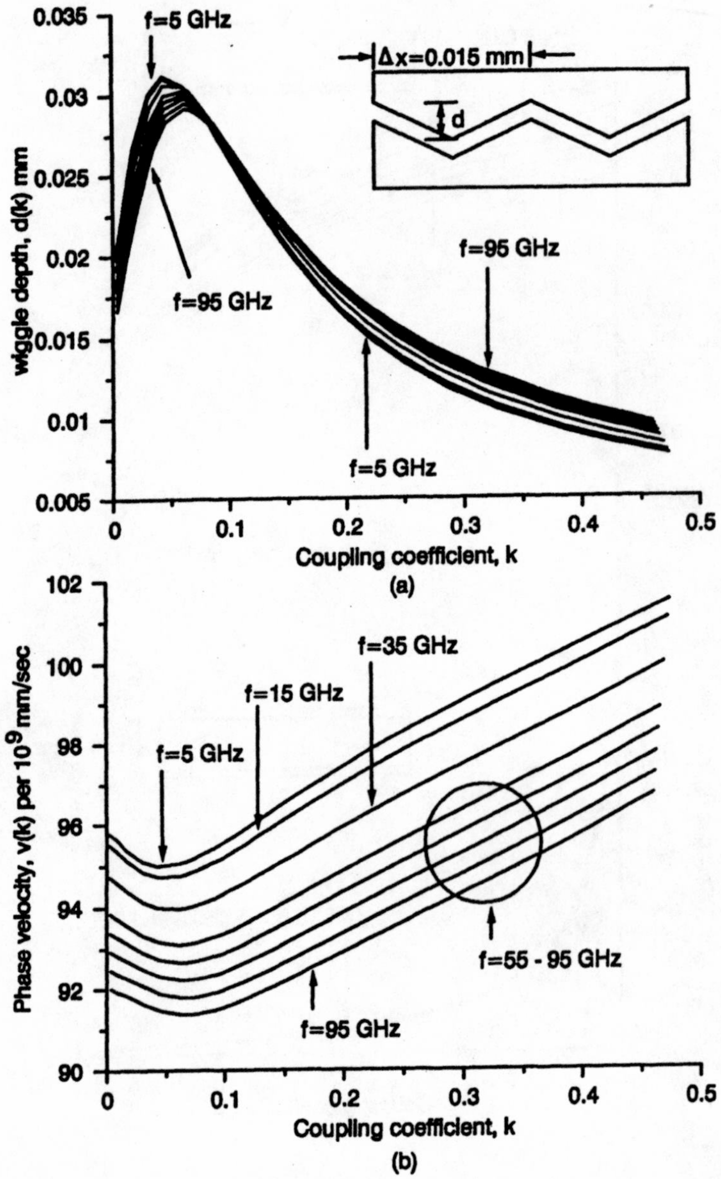

Figure 3.16. Design curves for double-coupled lines on GaAs substrate with $\epsilon_r = 12.9$, $h = 0.1$ mm, and $f = 5$–95 GHz: (a) wiggle depth $d(k)$ and (b) compensated phase velocity $v(k)$.

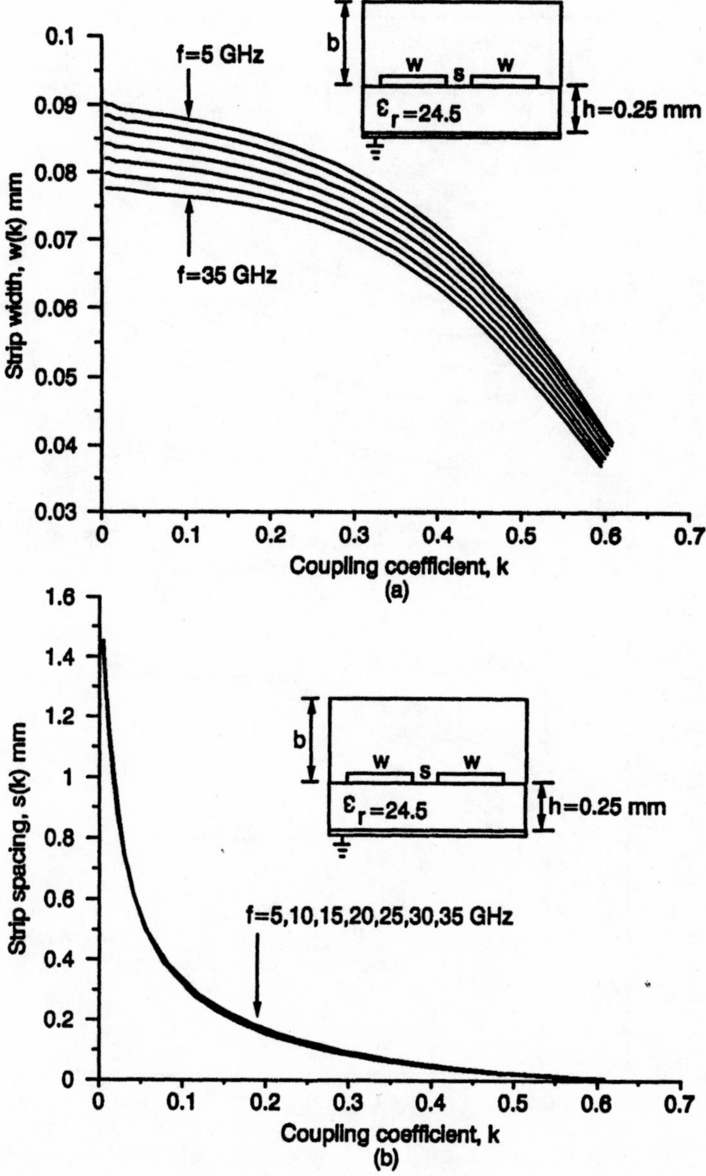

Figure 3.17. Design curves for double-coupled lines on Lanthalum Gallate substrate with $\epsilon_r = 24.5$, $h = 0.25$ mm, and $f = 5$–35 GHz: (a) strip width $w(k)$ and (b) strip spacing $s(k)$.

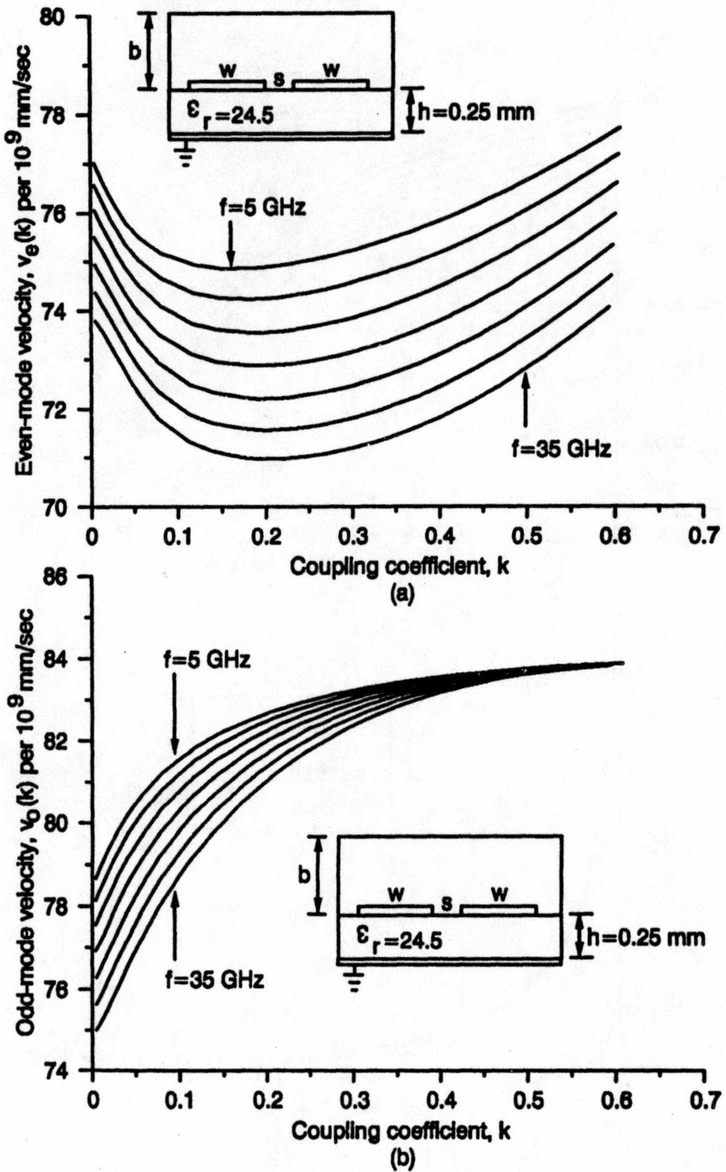

Figure 3.18. Design curves for double-coupled lines on Lanthalum Gallate substrate with $\epsilon_r = 24.5$, $h = 0.25$ mm, and $f = 5\text{–}35$ GHz: (a) even-mode phase velocity $v_e(k)$ and (b) odd-mode phase velocity $v_o(k)$.

form codirectional couplers with smooth coupled edges can advantageously be employed as they can be realized with large gaps.

3.4.9 Design Curves for Barium Tetratitanate Substrate with ϵ_r = 37 and h = 0.635 mm

The design curves for Barium Tetratitanate substrate are computed up to 20 GHz in steps of 5 GHz. The functions for strip width and spacing are given in Figure 3.19. The wiggle depth for phase velocity compensation is given in Figure 3.20. This substrate offers significant length reduction in nonuniform directional couplers and filters because of its very high relative dielectric constant. The $BaTi_4O_9$ substrate is reported to have low loss (0.5 dB at 10 GHz with a 26.1-Ω line on 0.381-mm thick substrate) and low dispersion [7].

3.4.10 Design Curves for RT-Duroid Substrate with ϵ_r = 10.5 and h = 0.2 mm

RT-Duroid is a soft substrate whose cutting and machining is very easy compared to ceramic substrates. It is usually used in applications requiring low relative dielectric constants (e.g., wideband microstrip antennas). Ceramic-filled RT-Duroid has high relative dielectric constant (ϵ_r = 10.5) and can be used for applications requiring via holes to ground (directional couplers used in phase shifters and balanced mixers). The design curves for this substrate are computed up to 35 GHz in steps of 5 GHz and are shown in Figures 3.21 and 3.22.

3.5 SYNTHESIS FUNCTIONS FOR NONUNIFORM COUPLED LINES

Synthesis of nonuniform coupled lines is carried out in two steps: synthesis of performance and synthesis of continuous physical dimensions. The former requires the compensated phase velocity at the design center frequency. In Section 1.6 we derived the coupled-arm response for the nonuniform directional coupler as

$$C(\omega) = \tanh\left\{\int_0^l \sin(2\omega x/v)p(x)\,dx\right\} \tag{3.13}$$

where $p(x)$ is the reflection coefficient distribution function given by

$$p(x) = \frac{1}{2}\frac{d}{dx}\ln[Z_{0e}(x)] \tag{3.14}$$

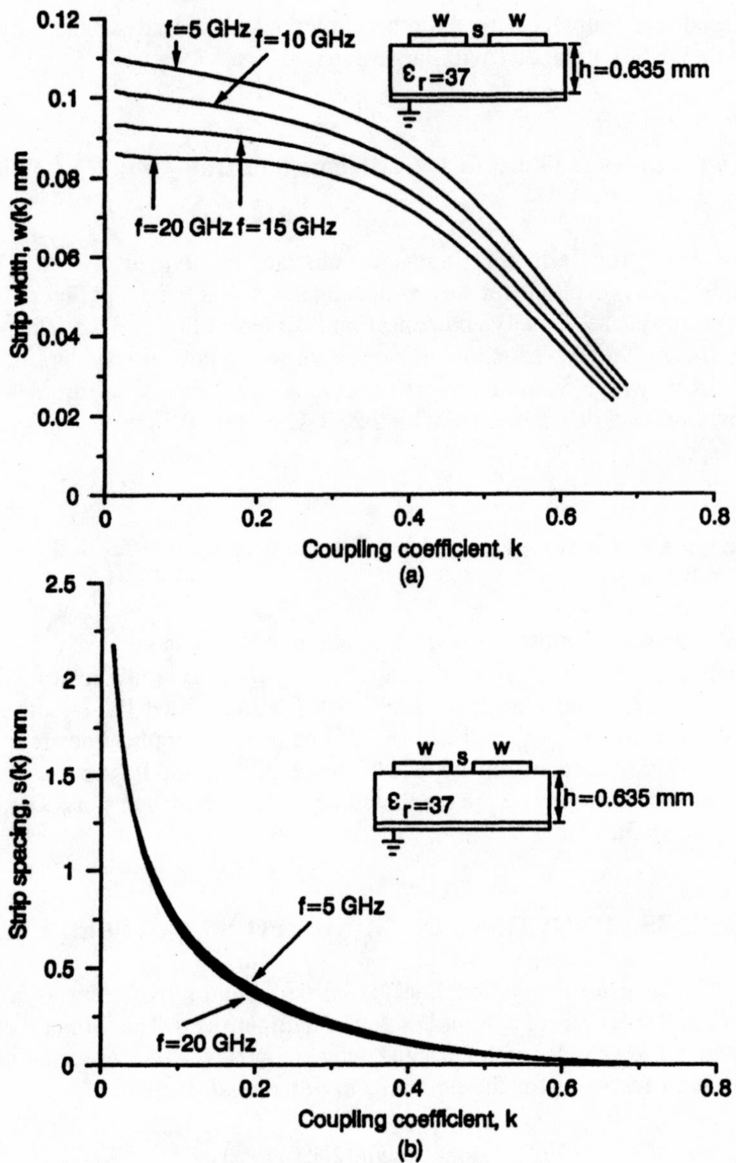

Figure 3.19. Design curves for double-coupled lines on Barium Tetratitanate substrate with $\epsilon_r = 37$, $h = 0.635$ mm, and $f = 5$–20 GHz: (a) strip width $w(k)$ and (b) strip spacing $s(k)$.

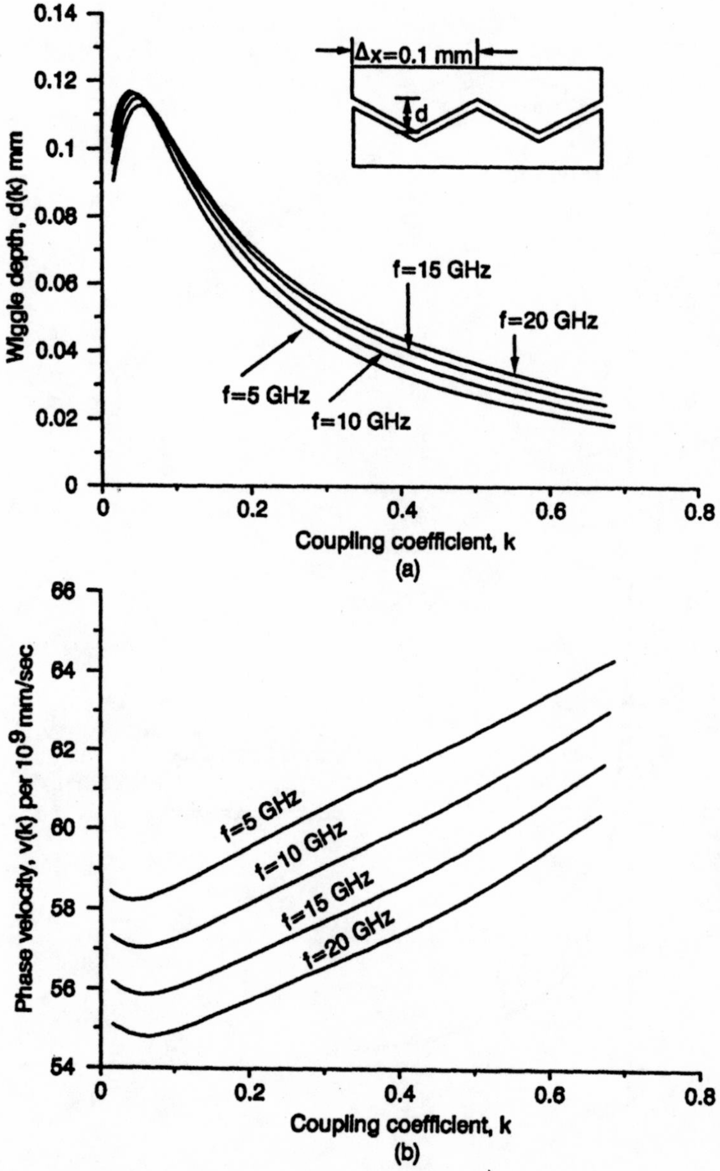

Figure 3.20. Design curves for double-coupled lines on Barium Tetratitanate substrate with $\epsilon_r = 37$, $h = 0.635$ mm, and $f = 5$–20 GHz: (a) wiggle depth $d(k)$ and (b) compensated phase velocity $v(k)$.

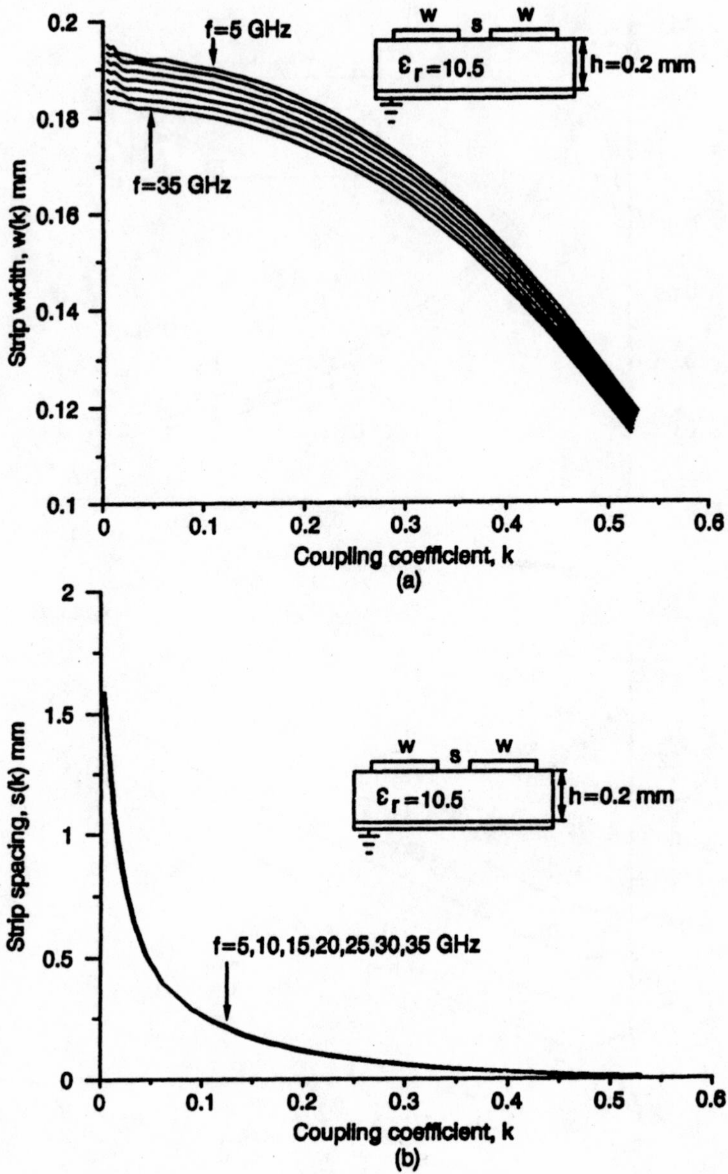

Figure 3.21. Design curves for double-coupled lines on 2 RT-Duroid substrate with $\epsilon_r = 10.5$, $h = 0.2$ mm, and $f = 5$–35 GHz: (a) strip width $w(k)$ and (b) strip spacing $s(k)$.

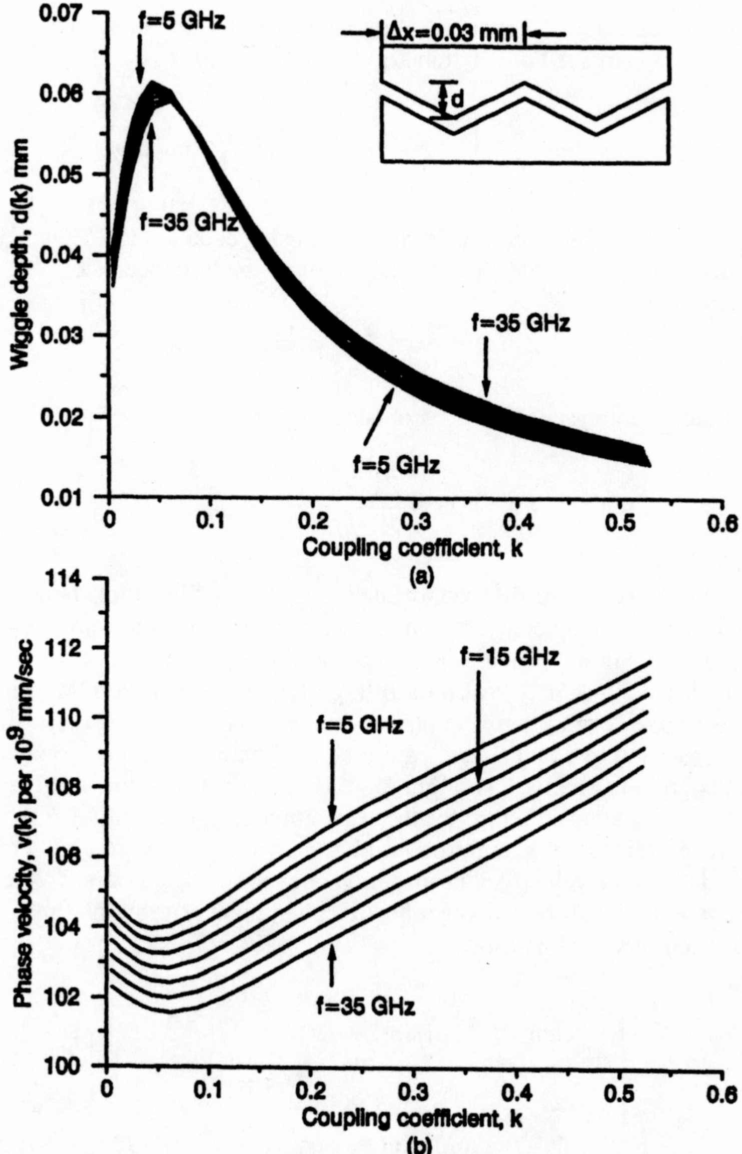

Figure 3.22. Design curves for double-coupled lines on 2 RT-Duroid substrate with $\epsilon_r = 10.5$, $h = 0.2$ mm, and $f = 5–35$ GHz: (a) wiggle depth $d(k)$ and (b) compensated phase velocity $v(k)$.

where $Z_{0e}(x)$ is the normalized even-mode impedance of the nonuniform coupled lines. The function $p(x)$ forms a Fourier transformation pair with $C(\omega)$ [8]:

$$p(x) = -\frac{2}{\pi v}\int_0^{2\omega_c} \sin(2\omega x/v)\tanh^{-1}[C(\omega)]\,d\omega \tag{3.15}$$

For a desired coupling function, $p(x)$ is computed using equation (3.15), which is then used in equation (3.14) to find the normalized even-mode impedance:

$$Z_{0e}(x) = \exp\left[2\int_0^x p(x)\,dx\right] \tag{3.16}$$

The continuous coupling coefficient is obtained by

$$k(x) = \frac{Z_{0e}^2(x) - 1}{Z_{0e}^2(x) + 1} \tag{3.17}$$

Very tight coupling (e.g., -3 dB) cannot be realized by double-coupled lines owing to the extremely small conductor separation. Tandem connection of loosely coupled lines in obtaining tight coupling provides a partial solution because the coupling values at the center of the coupler may still be too large for realization with double-coupled lines. Connection of several loose couplers in tandem may not be so practical because of increased insertion loss and space requirements. Therefore, whenever possible the designer opts for the smallest configuration. A practical solution is to employ a nonuniform interdigitated structure for the realization of tight coupling at the coupler center. However, this is not straightforward because introduction of an interdigitated section would affect the velocity of propagation of even and odd modes. To include the effect of the interdigitated section, the reflection coefficient distribution function given by (3.15) is then modified as follows:

$$p_{e,o}(x) = -\frac{2}{\pi v}\int_0^{2\omega_c} \tanh^{-1}[C(\omega)]\sin(2\omega x/v)\,d\omega \qquad -l/2 \le |x| \le -l_4/2 \tag{3.18}$$

$$p_{e,o}(x) = -\frac{2}{\pi v_{e,o}}\int_0^{2\omega_c} \tanh^{-1}[C_{e,o}(\omega)]\sin(2\omega x/v_{e,o})\,d\omega \qquad -l_4/2 \le x \le l_4/2 \tag{3.19}$$

where the subscripts e and o denote the even and odd modes, respectively; $C_{e,o}(\omega)$ is the respective coupling functions; l_4 is the length of the interdigitated section; and v_e and v_o are the phase velocities for the center section. The required $p(x)$ is then obtained by the superposition of the two modes:

$$p(x) = \frac{1}{2}[p_e(x) - p_o(x)] \quad (3.20)$$

At the midband, the design length for double-coupled lines is given by

$$l_{2c} = \frac{v}{4f_c} \quad (3.21)$$

where v is the compensated phase velocity and f_c is the design center frequency.

For the interdigitated coupler, the design length is

$$l_{4c} = \frac{v_e + v_o}{8f_c} \quad (3.22)$$

The overall nonuniform coupler length with interdigitated center section can then be written as

$$l = l_{2c}(n - q) + ql_{4c} \quad (3.23)$$

where n is the number of sections (which may be any value—it need not be an integer), and q is an adjustable variable that determines the total length for the interdigitated center section.

We have introduced the variable q for two reasons: (1) we would like to ensure a smooth transition from double-coupled lines to interdigitated section with minimum external modification and (2) we do not want unnecessary length of interdigitated section because of increased path length for bond wires at the transitions. The value of q is usually specified as between 0.5 and 1.

The second step in the synthesis of nonuniform coupled lines is the evaluation of continuous physical dimensions, which was briefly explained in Section 3.1. This step will become clearer in the next chapter, where we give design examples.

REFERENCES

[1] Sobhy, M. I., and E. A. Hosny, "The Design of Directional Couplers Using Exponential Lines in Inhomogeneous Media," *IEEE Trans. on Microwave Theory and Tech.*, Vol. MTT-30, 1982, pp. 71–76.

[2] Churchhouse, R. F., ed., *Handbook of Applicable Mathematics*, Vol. 3 of *Numerical Methods*, New York: John Wiley & Sons, 1981, pp. 234–241.

[3] Podell, A., "A High Directivity Microstrip Coupler Technique," *IEEE Trans. on Microwave Theory and Tech.*, 1970, pp. 33–36.

[4] Uysal, S., and A. H. Aghvami, "Synthesis, Design, and Construction of Ultra-Wide-Band Nonuniform Quadrature Directional Couplers in Inhomogeneous Media," *IEEE Trans. on Microwave Theory and Tech.*, Vol. MTT-37, 1989, pp. 969–976.

[5] Watkins, J., "The Onset of Anomalous Wave Generation in Patch Antennas," *Int. J. Electronics,* Vol. 65, 1988, pp. 849–864.

[6] Ganguly, A. K., and C. M. Krowne, "Characteristics of Microstrip Transmission Lines with High-Dielectric-Constant Substrates," *IEEE Trans. on Microwave Theory and Tech.,* Vol. MTT-39, 1991, pp. 1329–1337.

[7] Lee, Y. S., W. J. Getsinger, and L. R. Sparrow, "Barium Tetratitanate MIC Technology," *IEEE Trans. on Microwave Theory and Tech.,* Vol. MTT-27, 1979, pp. 655–660.

[8] Kammler, D. W., "The Design of Discrete *N*-Section and Continuously Tapered Symmetrical Microwave TEM Directional Couplers," *IEEE Trans. on Microwave Theory and Tech.,* Vol. MTT-17, 1969, pp. 577–590.

Chapter 4
Design and Construction of Ultrawideband Directional Couplers

4.1 INTRODUCTION

So far, we have discussed the synthesis of ultrawideband directional couplers. In this chapter we shall discuss their physical implementation using microstrip technology. Several design examples are presented, and whenever possible, verified measured results will also be given. Collected in the next section are the design steps for ultrawideband nonuniform directional couplers.

4.2 DESIGN PROCEDURE

A step-by-step design procedure follows.

Uniform Coupler Parameters

1. Input parameters.
   a. Select substrate: ϵ_r, h.
   b. Select characteristic impedance: Z_0.
   c. Set design frequency: f.
2. Consider uniform parallel-coupled lines.
   a. Set number of conductors: $N = 2$.
3. Specify strip separation: s.
4. Depending on input parameters, set range for strip width:

$$w_1 \leq w \leq w_2$$

5. Calculate quasistatic capacitances with $N = 2$, s, and $w_1 \leq w \leq w_2$: C_p, C_f, C_{fe}, C_{fo}, and C_{pa}, C_{fa}, C_{fea}, C_{foa}.

6. Calculate frequency-dependent effective dielectric constants at a given frequency: $\epsilon_{re}(f)$, $\epsilon_{ro}(f)$.
7. Optimize:

$$Z_0^2 c^2 C_{ea} C_{oa} \sqrt{\epsilon_{re}(f)\epsilon_{ro}(f)} - 1 = 0$$

   a. Obtain w.
8. Use specified s and optimized w to calculate frequency-dependent.
   a. Capacitances: C_p, C_f, C_{fe}, C_{fo}.
   b. Effective dielectric constants:

$$\epsilon_{re}(f) = \frac{C_e}{C_{ea}} \qquad \epsilon_{ro}(f) = \frac{C_o}{C_{oa}}$$

   c. Characteristic impedances:

$$Z_{0e}(f) = \frac{\sqrt{\epsilon_{re}(f)}}{cC_e} \qquad Z_{0o}(f) = \frac{\sqrt{\epsilon_{ro}(f)}}{cC_o}$$

   d. Phase velocities:

$$v_e(f) = \frac{c}{\sqrt{\epsilon_{re}(f)}} \qquad v_o(f) = \frac{c}{\sqrt{\epsilon_{ro}(f)}}$$

   e. Coupling coefficient:

$$k(f) = \frac{Z_{0e}(f) - Z_{0o}(f)}{Z_{0e}(f) + Z_{0o}(f)}$$

   f. Odd-mode capacitance parameter with wiggle:

$$C'_{fo} = (C_p + C_f) \times \left(\frac{\epsilon_{re}(f)}{\epsilon_{ro}(f)} - 1\right) + \frac{\epsilon_{re}(f)}{\epsilon_{ro}(f)} C_{fo}$$

9. Specify wiggle length: Δx.
10. Calculate wiggle depth:

$$d = \frac{\Delta x}{2} \exp\left[1.38 \frac{C_{fe}}{C_{fo}}\right] \sqrt{\left(\frac{C'_{fo}}{C_{fo}}\right)^2 - 1}$$

11. Calculate compensated phase velocity:

$$v = v_e/\exp\left[0.06\,\frac{C_{fe}}{C_{fo}}\right]$$

12. Repeat steps 3 to 11 with different values of f.
13. Repeat steps 3 to 12 with different values of s.
14. Output design parameters in tabular format: Table 1 column heads are k w s d f. Table 2 column heads are k v_e v_o v f.
15. Generate cubic splines (e.g., see Figures 3.3–3.22).
16. Set number of conductors: $N = 4$.
17. Repeat steps 3 to 15: with 8(first part) replaced by

$$k(f) = [Z_{0e}(f) - Z_{0o}(f)]/\left[Z_{0e}(f) + Z_{0o}(f) - \frac{2Z_{0e}Z_{0o}}{3(Z_{0e} + Z_{0o})}\right]$$

Omit steps 8 (last part) to 11.

Nonuniform Coupler Parameters

18. Specify desired coupling amplitude in the design bandwidth:

$$|C(\omega)|_s, B_w = f_1 - f_2$$

19. Specify allowable coupling ripple:

$$|C(\omega)| \pm r$$

20. Use compensated phase velocity at the design center frequency (step 11).
21. Specify an initial coupler length: l in millimeters.
22. Divide coupler length into elemental lengths: Δl.
23. Choose a convenient step size for angular frequency: $\Delta\omega$.
24. Evaluate the reflection coefficient distribution function from $x = -l/2$ to $x = 0$ in steps of Δl:

$$p(x) = -\frac{2}{\pi v}\int_0^{2\omega_c} \sin(2\omega x/v)\tanh^{-1}[C(\omega)]\,d\omega$$

25. Compute coupling response:

$$C(\omega) = \tanh\{\int_{-l/2}^{l/2} \sin(2\omega x/v)p(x)\,dx\}$$

26. Does specified length give desired bandwidth with allowable coupling ripple?
    a. If Yes, then specify error function in the design bandwidth:

$$C(\omega)_{err} = |C(\omega)|_s - C(\omega)$$

    and calculate the new input coupling function:

$$|C(\omega)|_{s_{new}} = |C\omega)|_s + C(\omega)_{err} \text{ or } |C(\omega)|_{s_{new}} = 2|C(\omega)|_s - C(\omega)$$

    Then repeat steps 24 to 25.
    b. If No, repeat steps 21 to 26 (first part) with different lengths l.
27. Compute the normalized even-mode impedance function:

$$Z_{0e}(x) = \exp[2 \int_0^x p(x)\,dx]$$

28. Compute the continuous coupling coefficient:

$$k(x) = \frac{Z_{0e}^2(x) - 1}{Z_{0e}^2(x) + 1}$$

29. Find the maximum value of $k(x)$ that occurs at the coupler center: $k(0)$. Is it realizable with double-coupled ($N = 2$) lines?
    a. If no, is it realizable with $N = 4$?
    b. If yes, determine the range of values for $k(x)$ that are not realizable with $N = 2$, that is, $k(-l_a)$ to $k(l_a)$ and introduce interdigitated coupler section for the following values.
    c. Calculate v_e and v_o (with $N = 4$) at the design center frequency. Calculate the interdigitated coupler length: $l_4 = (v_e + v_o)/8f_c$.
    d. Is $l_4 = 2l_a$? If yes, go to step 30. If no, calculate q: $q = 2l_a/l_4$ and go to step 30.
    e. If no, use tandem configuration. Calculate the new input coupling for each coupler in tandem

$$|C(\omega)|_{s_{tandem}} = \sin\left[\frac{\sin^{-1}(|C(\omega)|_s)}{2}\right]$$

    and repeat steps 24 onward.
    f. If yes, produce values of $k(x)$ from $-l/2$ to 0 (as we have symmetrical couplers) and go to step 31.
30. Calculate new values of $p(x)$:

$$p_{e,o}(x) = -\frac{2}{\pi v}\int_0^{2\omega_c} \tanh^{-1}[C(\omega)]\sin(2\omega x/v)\,d\omega \quad -l/2 \le |x| \le -ql_4/2$$

$$p_{e,o}(x) = -\frac{2}{\pi v_{e,o}}\int_0^{2\omega_c} \tanh^{-1}[C(\omega)]\sin(2\omega x/v_{e,o})\,d\omega \quad -ql_4/2 \le x \le ql_4/2$$

a. Repeat steps 25 to 29.

31. For $N = 2$, calculate $w(x)$, $s(x)$ and $d(x)$ by evaluating $w(k)$, $s(k)$ and $d(k)$, respectively, at values of $k(x)$ from $-l/2$ to $-ql_4/2$. For $N = 4$, calculate $w(x)$ and $s(x)$ by evaluating $w(k)$ and $s(k)$, respectively, at values of $k(x)$ from $-ql_4/2$ to 0.
32. Tabulate the data as follows:

N_2 Δl
x $w(x)$ $s(x)$ $d(x)$
N_4 Δl
x $w(x)$ $s(x)$

(Note: $N_2\,\Delta l + N_4\,\Delta l = l$.)

33. Generate a coupler layout (see the next section).
34. Build the coupler.
35. Measure its performance, Is the performance acceptable?
    a. If yes, congratulations.
    b. If no, check manufacturing tolerances. If you are not satisfied rebuild the coupler by redefining tolerances.
37. The most probable source of error is in the selection of phase velocities. These, in general, are functions of both frequency and the coupling coefficient, as can be seen from the computed design curves given in Chapter 3. However, in many cases, variation in phase velocities with frequency can be taken as linear. Therefore, as far as the frequency is concerned, a median value can still be selected. The overall effect is an increase in ripple amplitude. This can be corrected by specifying less ripple amplitude in the design bandwidth. To determine the new ripple amplitude,
    a. Compute $p(x)$ with velocities at f_c.
    b. Compute $C(\omega)$ with $p(x)$ but replace v with its minimum value (lower band edge) and then its maximum value (upper band edge).

Except in designs with a relatively large ripple amplitude, the variation in velocity along the coupler length cannot be neglected. This is because, with ultrawideband designs having a relatively longer length, a significant portion of the double-coupled region is weakly coupled. Due to this weak coupling, larger wiggle depth is required, thereby lowering the even-mode phase velocity causing a nonlinear variation along the coupler length. Therefore, at this stage, the synthesized $k(x)$ should be used to compute $\upsilon(x)$ by evaluating cubic spline function $\upsilon(k)$. The coupling function is then computed by

$$C(\omega) = \tanh\left\{\int_0^l \sin\left(2\omega \int_0^x \frac{dx}{\upsilon(x)}\right) p(x)\, dx\right\}$$

4.3 NONUNIFORM COUPLER LAYOUT

The nonuniform coupler physical dimensions are defined by continuous functions for width, spacing, and wiggle depth as $w(x)$, $s(x)$, and $d(x)$, respectively. These functions correspond to continuous coupling coefficient $k(x)$, which is synthesized and obtained by a suitable choice of elemental coupler length. Therefore, it is very critical that the physical layout of the coupler represent these functions as accurately as possible [1].

In general, the nonuniform coupler geometry can be calculated by generating equations defining the inner and outer edges of double-coupled lines, wiggly inner edges, and interdigitated center section. First, consider the nonuniform double-coupled lines coordinate system shown in Figure 4.1. Because the coupler is symmetrical, it is sufficient to consider, say, the upper left quarter only. The coupler is enclosed in a rectangle (mask area) whose bottom left corner is taken as the origin for the coupler coordinate system.

The parameters w_1 and s_1 shown in this figure denote initial strip width and spacing at $k(-l/2)$, respectively. The equations y_{11} and y_{12} define the inner and outer

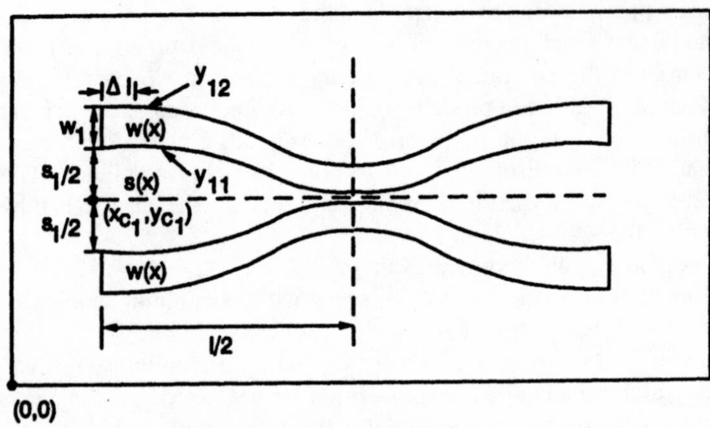

Figure 4.1 Nonuniform coupled lines coordinate system.

edges of the coupled line (line 1), respectively. The lower part of the coupler (line 2) is the mirror image of the top part about the line $y = y_{c_1}$.

Let us assume that the axial length $l/2$ is divided into n elemental sections: $\Delta l = (l/2)/n$. The elemental section Δl is shown in Figure 4.2. With a small Δl, we can approximate the curves y_{11} and y_{12} by linear functions again denoted by y_{11} and y_{12}, which are shown as dashed lines. Using the equation for a line, $y = mx + c$, we can define the inner and outer edges by the following equations:

$$y_{11_1} = x_{c_1} q_{11_1} + y_{c_1} + \frac{s_1}{2} \tag{4.1}$$

$$y_{12_1} = x_{c_1} q_{12_1} + y_{c_1} + w_1 + \frac{s_1}{2} \tag{4.2}$$

where q_{11_1} and q_{12_1} are the slopes for the inner and outer edges, respectively, given by

$$q_{11_1} = \frac{s_2 - s_1}{2\Delta l} \tag{4.3}$$

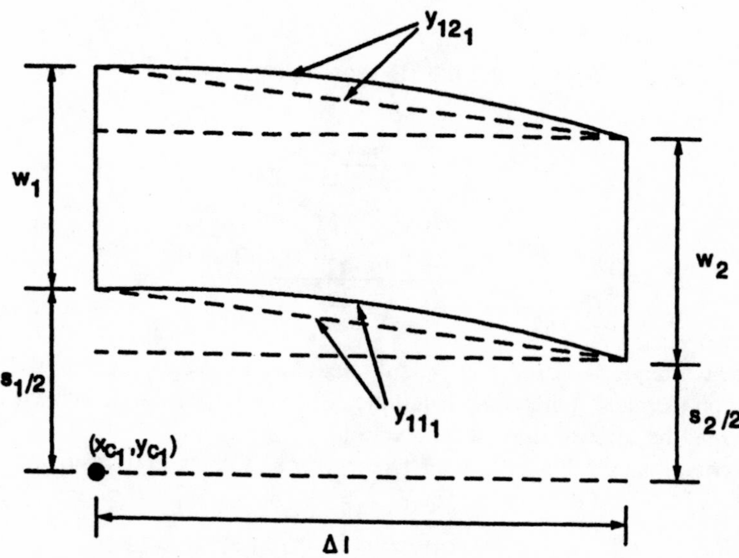

Figure 4.2 Elemental coupler length and its associated parameters.

$$q_{12_1} = \frac{s_2/2 + w_2 - (s_1/2 + w_1)}{\Delta l} \tag{4.4}$$

For the next elemental length we have

$$y_{11_2} = x_{c_2} q_{11_2} + y_{c_1} + \frac{s_2}{2} \tag{4.5}$$

$$y_{12_2} = x_{c_2} q_{12_2} + y_{c_1} + w_2 + \frac{s_2}{2} \tag{4.6}$$

where $x_{c_2} = x_{c_1} + \Delta l$ and q_{11_2} and q_{12_2} are the slopes for the inner and outer edges of the coupled line for the second elemental length.

Generalizing these equations, we obtain

$$y_{11_i} = x_{c_i} q_{11_i} + y_{c_1} + \frac{s_i}{2} \tag{4.7}$$

$$y_{12_i} = x_{c_i} q_{12_i} + y_{c_1} + w_i + \frac{s_i}{2} \tag{4.8}$$

where $i = 1, 2, \ldots, n - 1$, and the slopes are given by

$$q_{11_i} = \frac{s_{i+1} - s_i}{2\Delta l} \tag{4.9}$$

$$q_{12_i} = \frac{s_{i+1}/2 + w_{i+1} - (s_i/2 + w_i)}{\Delta l} \tag{4.10}$$

At microwave frequencies, it is usually sufficient to choose $\Delta l = 0.05$ mm. However, depending on the mask generation technique, a smaller value can be chosen. Alternatively, Δl may be divided into Δl_s subsections [1].

The equations for the second conductor may be written in terms of y_{11_i} and y_{12_i}:

$$y_{21_i} = 2y_{c_1} - y_{11_i} \tag{4.11}$$

$$y_{22_i} = 2y_{c_1} - y_{12_i} \tag{4.12}$$

The next step is to introduce wiggling into the coupled region. The wiggly section is shown in Figure 4.3. The axial length, Δx, of each wiggle section is subdivided into m sections. Each subdivided section has a length Δl, which is the same elemental length used in the previous equations. The introduction of wiggle must not disturb the total capacitance parameter C_p. At this state, we first decide on a convenient value for Δx. We also prefer that the number of subdivisions be an even number. The other restriction that we impose on Δx is such that $d_c/2 \leq w/2$ at that point.

The initial value of wiggle depth is computed first

$$A_{d_o} = \frac{d_1 + \ldots + d_{m/2}}{2} \tag{4.13}$$

The wiggle slope is given by

$$|q_{11w_i}| = \frac{d_i}{\Delta l} \tag{4.14}$$

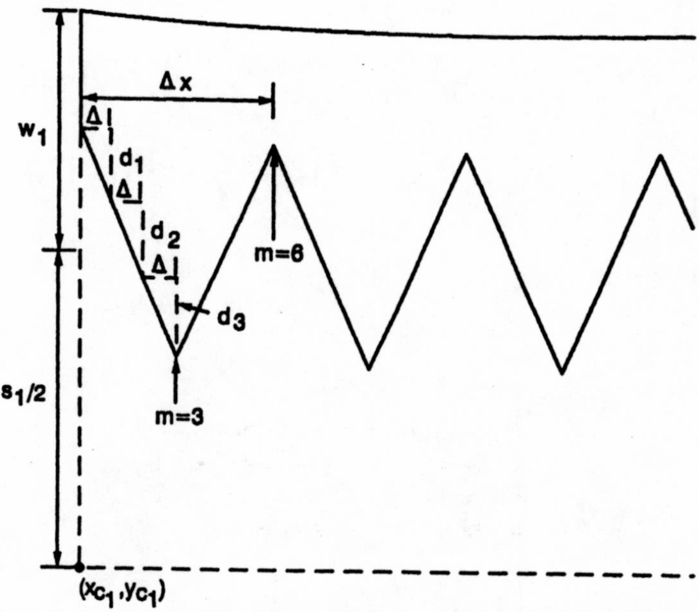

Figure 4.3 Introduction of wiggle into the coupled inner edge of a nonuniform line.

With the wiggle, (4.7) becomes

$$y_{11w_i} = \begin{cases} y_{11_i} + A_{d_i} + (x_{a_i} - x_{b_i})q_{11w_i}, & i = \left[1,\ldots,\frac{m}{2}\right], \\ & \left[(m+1),\ldots,\left(m+\frac{m}{2}\right)\right],\ldots \\ \\ y_{11_i} + A_{d_i} + (x_{b_i} - x_{m_i})q_{11w_i}, & i = \left[\left(\frac{m}{2}+1\right),\ldots,m\right], \\ & \left[\left(m+\frac{m}{2}+1\right),\ldots,(2m)\right],\ldots \end{cases} \tag{4.15}$$

where

$$A_{d_i} = \begin{cases} A_{d_0} & i = 1,\ldots,\frac{m}{2} \\ 0.5[x_{a_i} - x_{b_{(m/2)}}]q_{11w_i}, & i = \left(\frac{m}{2}+1\right),\ldots,\left(m+\frac{m}{2}\right) \\ 0.5[x_{a_i} - x_{b_{(m+m/2)}}]q_{11w_i}, & i = \left(m+\frac{m}{2}+1\right),\ldots,\left(2m+\frac{m}{2}\right) \\ 0.5[x_{a_i} - x_{b_{(2m+m/2)}}]q_{11w_i}, & i = \left(2m+\frac{m}{2}+1\right),\ldots,\left(3m+\frac{m}{2}\right) \\ \cdot & \cdot \\ \cdot & \cdot \\ \cdot & \cdot \end{cases}$$

$$x_{a_i} = \begin{cases} x_{b_1}, & i = 1,\ldots,m \\ x_{b_{(m+1)}}, & i = (m+1),\ldots,2m \\ x_{b_{(2m+1)}}, & i = (2m+1),\ldots,3m \\ \\ \cdot & \cdot \\ \cdot & \cdot \\ \cdot & \cdot \end{cases}$$

$$x_{m_i} = \begin{cases} x_{b_{(m/2)}}, & i = 1,\ldots,\left(m + \frac{m}{2} - 1\right) \\ x_{b_{(m+m/2)}}, & i = \left(m + \frac{m}{2}\right),\ldots,\left(2m + \frac{m}{2} - 1\right) \\ x_{b_{(2m+m/2)}}, & i = \left(2m + \frac{m}{2}\right),\ldots,\left(3m + \frac{m}{2} - 1\right) \\ \cdot & \cdot \\ \cdot & \cdot \\ \cdot & \cdot \end{cases}$$

$$x_{b_i} = x_{c_1} + \sum_{i=1} (i - 1)\Delta l$$

$$i = 1, 2, \ldots, \left(\frac{l/2}{\Delta l} - 1\right)$$

and y_{11_i} is given by equation (4.7). The value of y_{12_i} is not affected by wiggling; that is, $y_{12w_i} = y_{12_i}$. A similar formulation can be carried out for the second conductor:

$$y_{21w_i} = \begin{cases} y_{21_i} + A_{d_i} + (x_{a_i} - x_{b_i})q_{21w_i}, & i = \left[1,\ldots,\frac{m}{2}\right], \\ & \left[(m+1),\ldots,\left(m + \frac{m}{2}\right)\right],\ldots \\ y_{21_i} + A_{d_i} + (x_{b_i} - x_{m_i})q_{21w_i}, & i = \left[\left(\frac{m}{2} + 1\right),\ldots,m\right], \\ & \left[\left(m + \frac{m}{2} + 1\right),\ldots,(2m)\right],\ldots \end{cases} \tag{4.16}$$

where y_{21_i} is given by equation (4.11) and other parameters have the same value as previously.

4.3.1 Interdigitated Center Section

A possible way of connecting the double-coupled lines to the interdigitated center section is illustrated in Figure 4.4. In this case we need a total of four equations to

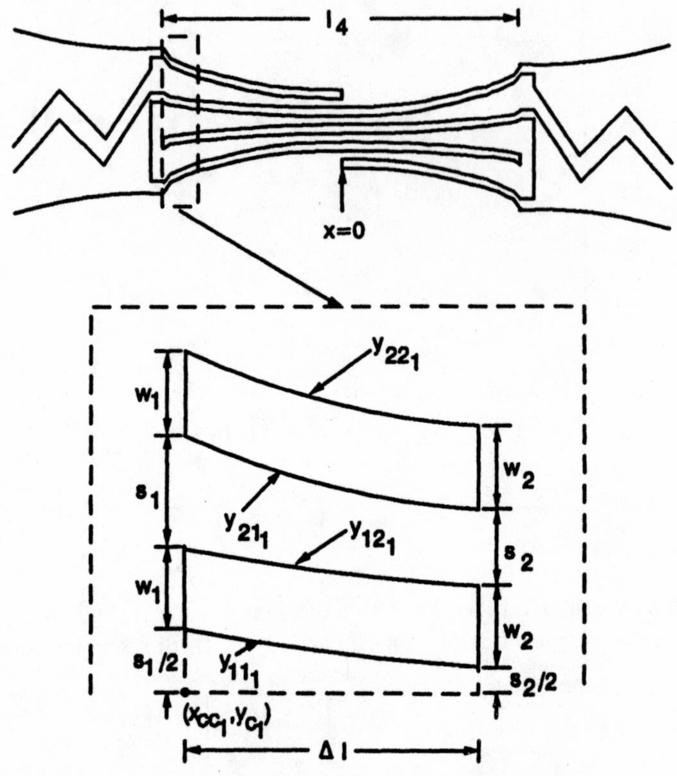

Figure 4.4 Transition form double-coupled lines to interdigitated center section.

define the geometry of the nonuniform interdigitated section. The coordinate (x_{cc_1}, y_{c_1}) is taken as the initial coordinate. The equations defining the two conductors can be derived in a way similar to double-coupled lines. Thus,

$$y_{11_i} = x_{cc_i} q_{11_i} + y_{c_1} + \frac{s_i}{2} \tag{4.17}$$

$$y_{12_i} = x_{cc_i} q_{12_i} + y_{c_1} + w_i + \frac{s_i}{2} \tag{4.18}$$

$$y_{21_i} = x_{cc_i} q_{21_i} + y_{c_1} + w_i + 3\frac{s_i}{2} \tag{4.19}$$

$$y_{22_i} = x_{cc_i} q_{22_i} + y_{c_1} + 2w_i + 3\frac{s_i}{2} \tag{4.20}$$

where $x_{cc_i} = x_{c_i} + l_2/2$, q_{11_i}, q_{12_i}, q_{21_i}, and q_{22_i} are the slopes for the respective lines and $i = 1, 2, \ldots, (n - 1)$ (one-half of the center section is divided into n subsections).

For the righthand side ($x > 0$), the preceding formulation is still valid, with y_{c_1} replaced by y_{cc_1}:

$$y_{cc_1} = y_{c_1} - (s_c + w_c)$$

where s_c and ω_c are the conductor spacing and width at $x = 0$. However, the transition from double-coupled lines to the interdigitated center section is not that straightforward. For a smooth transition we need

$$2w_2 + s_2 = 4w_4 + 3s_4$$

where w_2 and s_2 are the width and spacing for the double-coupled lines at $x = -l_4/2$, and w_4 and s_4 are the width and spacing for the interdigitated conductors at the same point. The minimum realizable s_2 usually dictates the length for the interdigitated section. That is, if $s_2 = 0.05$ mm (with ordinary photolithographic techniques), then an interdigitated center section is used for those coupling values that give $s_2 < .05$ with double-coupled lines.

4.4 DESIGN EXAMPLES

In this section, we give examples to highlight some of the possible approaches in ultrawideband coupler design using microstrip technology. Some measured results will also be given to verify the theory.

It is very difficult to determine the most practical value for conductor spacing that could be considered to correspond to zero coupling. This controversial situation can be avoided by setting the initial coupling to a finite but very small value. In the following design examples the even-mode impedance will be multiplied by a constant that will increase the initial coupling coefficient ($k(-l/2)$) to a desired finite value. However, this constant value causes an increase in the ripple amplitude of the coupling function, which will now be considered.

The normalized even-mode impedance computed at the design step 27 is multiplied by a constant so that the new impedance function is greater than 1:

$$Z_{0e_A}(x) = Z_{0e}(x)e^A \tag{4.21}$$

where A is a constant.

The new reflection coefficient distribution function is given by

$$p_A(x) = \frac{1}{2}\frac{d}{dx}\ln Z_{0e}(x) + \frac{1}{2}\frac{d}{dx}A \tag{4.22}$$

or

$$p_A(x) = p(x) + \frac{A}{2}\{\delta(x + l/2) - \delta(x - l/2)\} \tag{4.23}$$

The new coupling function is then obtained as

$$C_A(\omega) = C(\omega) + \frac{A}{2}\{e^{j\omega l/v} - e^{-j\omega l/v}\} \tag{4.24}$$

or

$$C_A(\omega) = C(\omega) + jA\sin(\omega l/v) \tag{4.25}$$

This is a key design feature that gives the design engineer more control on the design parameters that affect the taper rate of the coupled lines. As a result, the coupler can be squeezed in a smaller area because the initial spacing can be reduced to a desired value. In all the following design examples the value of A will be taken as 0.02.

4.4.1 2–20 GHz, −20 dB Coupler on Alumina Substrate with $\epsilon_r = 9.9$ and $h = 0.635$ mm

First, set the initial specifications: $Z_0 = 50\ \Omega$, $f_c = 10$ GHz, $N = 2$, $|C(\omega)|_s = 0.1$, $B_w = 2$–20 GHz, $r = \pm 1.5$ dB. Choose the phase velocity, an initial coupler length, and step sizes for ω and l: $v = 108 \times 10^9$ mm/s, $l = 5l_c = 13.5$ mm, $\Delta l = 0.05$ mm, and $\Delta\omega = 0.1$ Then follow design steps 24 to 33.

The computed reflection coefficient distribution function is shown in Figure 4.5(a). The corresponding continuous coupling coefficient is shown in Figure 4.5(b). Because the maximum value of $k(x)$ is 0.218, this function can be realized using double-coupled lines only. The function $k(x)$ is then used to evaluate the physical dimensions by using the design curves first discussed in Chapter 3. The continuous physical dimensions $w(x)$, $s(x)$, and $d(x)$ are shown in Figure 4.5(c). We see that there are two regions where the slopes of physical dimensions change sign (this is clearly visible from $s(x)$). This is a direct result of the optimization of the coupling function in order to achieve equal ripple performance in the design bandwidth. This can also be

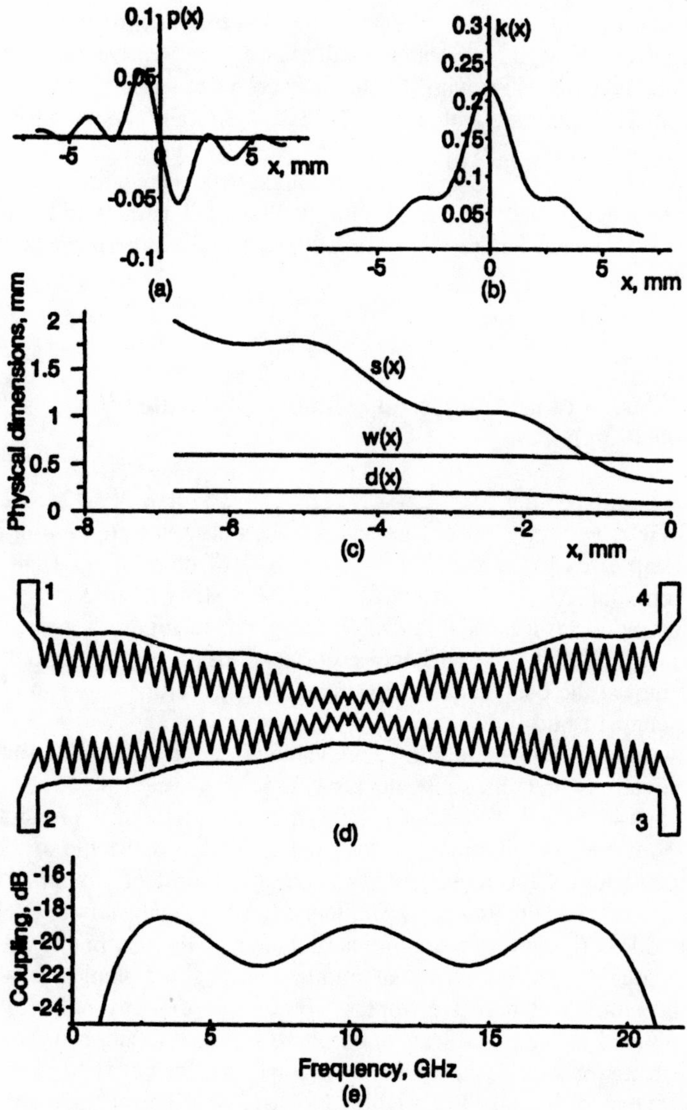

Figure 4.5 Computed results for the 2–20 GHz, −20 dB coupler on 0.635-mm-thick alumina substrate: (a) reflection coefficient distribution function, (b) continuous-coupling coefficient, (c) continuous physical dimensions, (d) coupler layout, and (e) coupling response.

seen from the computed reflection coefficient distribution function. At these regions, it becomes negative for negative x values and positive for positive x values.

The computed values for wiggle depth $d(x)$ correspond to an elemental length of $\Delta x = 0.1$ mm. The maximum value of $d(x)$ is 0.214 mm. For the nonuniform coupler we choose $\Delta x = 0.3$ mm. This gives a total of six elemental lengths ($\Delta l = 0.05$ mm) in each wiggle. Therefore, consecutive three wiggle depths are added on or subtracted from the inner edges of the conductors with the way as illustrated in Figure 4.3. The design layout is shown in Figure 4.5(d). Finally, the predicted performance is given in Figure 4.5(e).

4.4.2 2–20 GHz, −14 dB Coupler on Alumina Substrate with $\epsilon_r = 9.9$ and $h = 0.635$ mm

First, set the initial specifications: $Z_0 = 50\ \Omega$, $f_c = 10$ GHz, $N = 2$, $|C(\omega)|_s = 0.2$, $B_w = 2$–20 GHz, $r = \pm 1.5$ dB. Then choose the phase velocity, an initial coupler length, and step sizes for ω and l: $v = 108 \times 10^9$ mm/s, $l = 5l_c = 13.5$ mm, $\Delta l = 0.05$ mm, and $\Delta\omega = 0.1$. Then follow design steps 24 to 33.

The computed results for this design are given in Figure 4.6. The maximum value of $k(x)$, Figure 4.6(b), is 0.418, which corresponds to a spacing of 0.069 mm and width 0.432 mm at the coupler center. This coupler, therefore, can again be realized with double-coupled lines only.

The maximum nominal coupling achievable with double-coupled lines depends on the specifications and the substrate type. It is a common practice to limit the minimum value of spacing to about 0.05 mm with ordinary photolithographic techniques. However, on alumina substrate the minimum realizable spacing can be as low as 0.020 mm. This requires very accurate control of "external" parameters such as etchants and etching times and temperatures. Again, a fractionally smaller bandwidth and less ripple mean a significant change in the maximum value of $k(x)$. Because for equal ripple couplers the number of positive ripples is always one more than the number of negative ripples—Figure 4.6(c)—any reduction in ripple amplitude implies a lower $k(0)$. It can be observed from the function $s(k)$ of Chapter 3 that a small change in s causes a significant change in k for $k > 0.30$. It is, therefore, highly recommended that all the variables affecting $k(0)$ be properly accounted for when designing ultrawideband nonuniform directional couplers. The continuous wiggle depth for this design is computed in the same way as described in the previous design example. The design layout, which is not shown, is very similar to that of Figure 4.5(d) except that the conductor spacing tapers down to a smaller value at the coupler center. The computed coupling response for this design is shown in Figure 4.6(c).

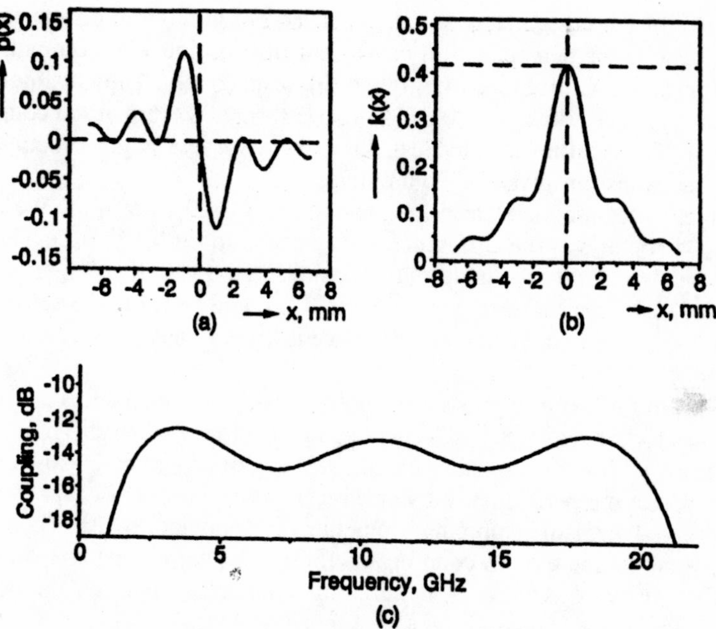

Figure 4.6 Computed results for the 2–20 GHz, −14 dB coupler on 0.635-mm thick alumina substrate: (a) reflection coefficient distribution function, (b) continuous-coupling coefficient, and (c) coupling response.

4.4.3 2–20 GHz, −3 dB Coupler on Alumina Substrate with $\epsilon_r = 9.9$ and $h = 0.635$ mm

First, set the initial specifications: $Z_0 = 50\ \Omega$, $f_c = 10$ GHz, $N = 2$, $|C(\omega)|_s = 0.707$, $B_w = 2$–20 GHz, $r = \pm 0.7$ dB. Then choose the phase velocity, an initial coupler length, and step sizes for ω and l: $v = 108 \times 10^9$ mm/s, $l = 5l_c = 13.5$ mm, $\Delta l = 0.05$ mm, and $\Delta\omega = 0.1$. Then follow design steps 24 to 33.

Design step 29 immediately reveals that the synthesized $k(0)$ is not realizable either with $N = 2$ or $N = 4$. So, we choose tandem configuration. The new input coupling is then obtained by

$$|C(\omega)|_{s_{\text{tandem}}} = \sin[\sin^{-1}(|C(\omega)|_s)/2] = 0.383$$

Design steps 24 onward are then repeated, and we find that we need an interdigitated ($N = 4$) center section for $-1.4 \le x \le 1.4$. For the center section, we choose $v_e = 116 \times 10^9$ mm/s and $v_o = 128 \times 10^9$ mm/s with $q = 0.9$.

The rest of the design steps are then carried out giving the results as shown in Figure 4.7. The computed reflection coefficient distribution function and the corresponding $k(x)$ are given in Figure 4.7(a) and (b), respectively. The maximum value of $k(x)$ is 0.66, which requires $w = 0.075$ mm and $s = 0.082$ mm at the coupler center with $N = 4$. The continuous physical dimensions for the wiggly sections and interdigitated sections are given in Figure 4.7(c).

We then calculate the transition values from $N = 2$ to $N = 4$. We found that $4w_4 + 3s_4 > w_2 + s_2$. The difference is computed to be 0.158 mm. This value is acceptable and we modify the widths and spacings in the vicinity of the transition (we slightly increase s_2 and w_2 at $x = -1.50$, -1.45, and -1.4 and reduce s_4 at $x = -1.4$, -1.35, and -1.3). When the difference is high, it may be necessary to change the value of q.

The layout for this design is shown in Figure 4.7(d). The lengths of the connecting 50 Ω lines for the two −8.34 dB couplers in tandem are taken as 3 mm. This value is by no means optimum. A smaller length, say, 2 mm or less can also be used. This would then affect the performance by causing small ripples on all four S-parameters due to increased parasitic coupling from the first coupler to the second coupler. However, when the increase in coupling coefficient is sharp at the coupler edges and the available space is scarce, a 1-mm long connecting line would do the job. Otherwise, this length should be kept at 3 mm.

From a given horizontal axis the length of the 50 Ω line connected to the coupled port is shorter than the 50 Ω line connected to the direct port. This difference is equal to $s_c + w_c$ (s_c and w_c are the spacing and width at $x = 0$). In this design, this value is 0.157 mm, which causes a linear phase variation in phase quadrature with a maximum of 10° at 20 GHz. Therefore, this value should be taken into account when designing tandem quadrature couplers.

This coupler is then built on $25.4 \times 25.4 \times 0.635$ mm^3 alumina substrate by changing the input and output, and the connecting 50 Ω lines into circular curves as shown in Figure 4.8.

The measurements were carried out on an HP8510C automatic network analyzer. The computed and measured coupling balance are given in Figure 4.9(a) and (b), respectively. Good agreement is observed in the entire bandwidth except that the measured coupling is slightly lower than the predicted ones. A close inspection of the realized dimensions revealed that the coupler was overetched by 0.008 mm.

The measured input reflection, Figure 4.9(c), is better than −15 dB in the design bandwidth. For the transitions, 3.5 mm dc-18 GHz OSM connectors were used. Therefore, above 18 GHz the measured results should be treated accordingly.

The measured isolation is around −15 dB and worsened slightly as a direct result of overetching, which is shown in Figure 4.9(d). Deviation from phase quadrature is less than ±3° up to 14 GHz, and ±5° up to 18 GHz; this is shown in Figure 4.9(e).

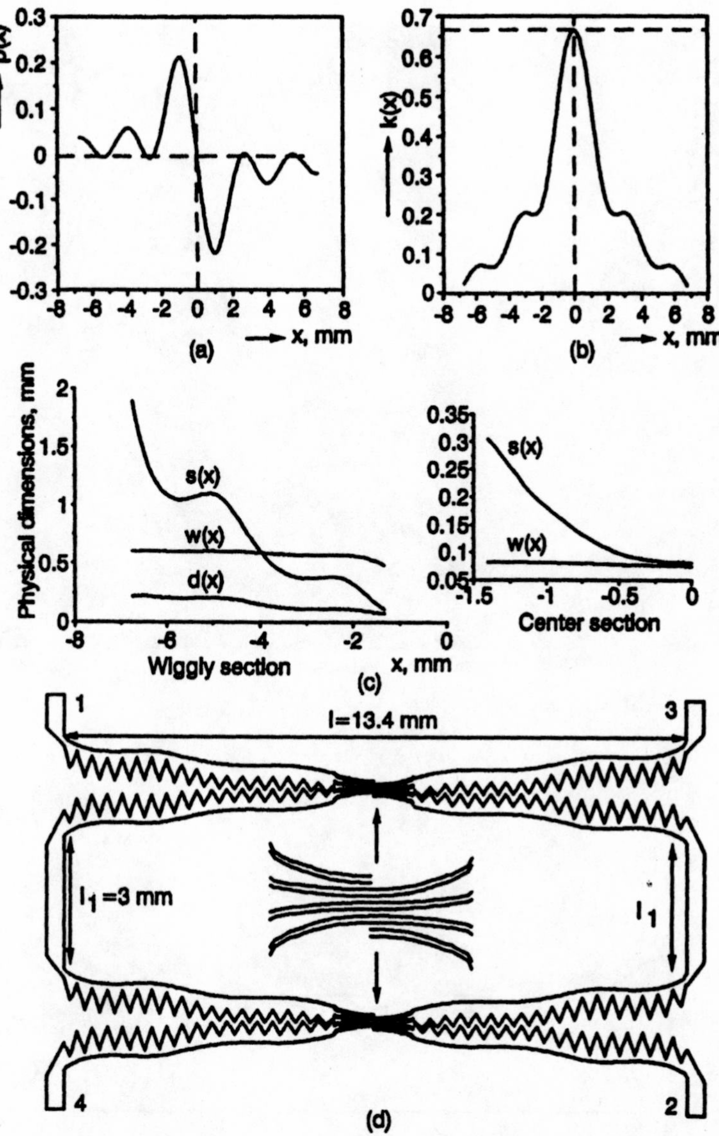

Figure 4.7 Computed results for the 2–20 GHz, −3 dB coupler on 0.635-mm-thick alumina substrate: (a) reflection coefficient distribution function, (b) continuous-coupling coefficient, (c) continuous physical dimensions, and (d) coupler layout.

Figure 4.8 The 2–20 GHz, −3 dB coupler.

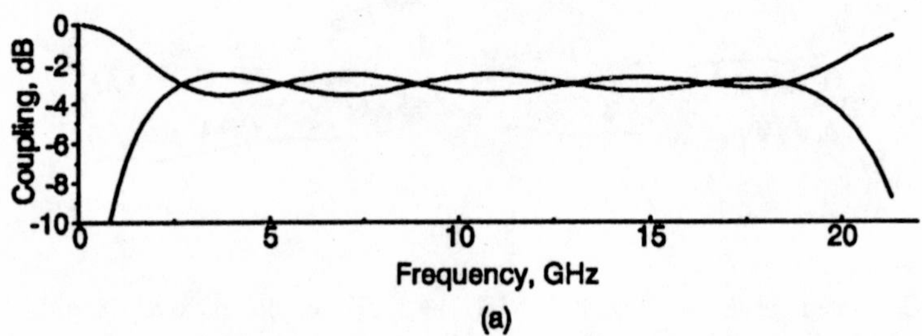

Figure 4.9 Performance for the 13.5-mm long −3 dB quadrature coupler on alumina substrate: (a) computed coupled and direct ports.

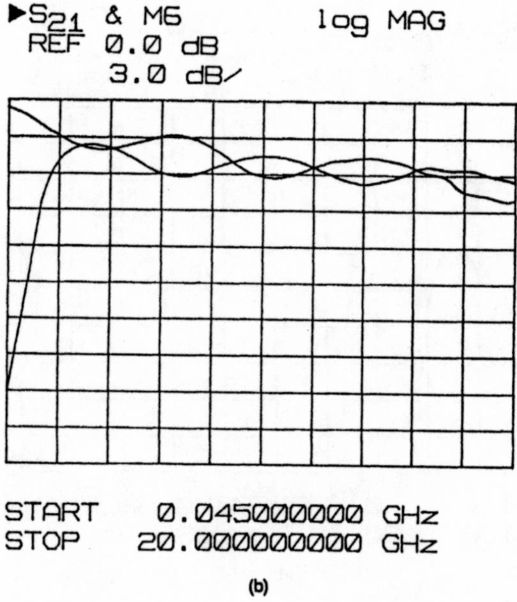

(b)

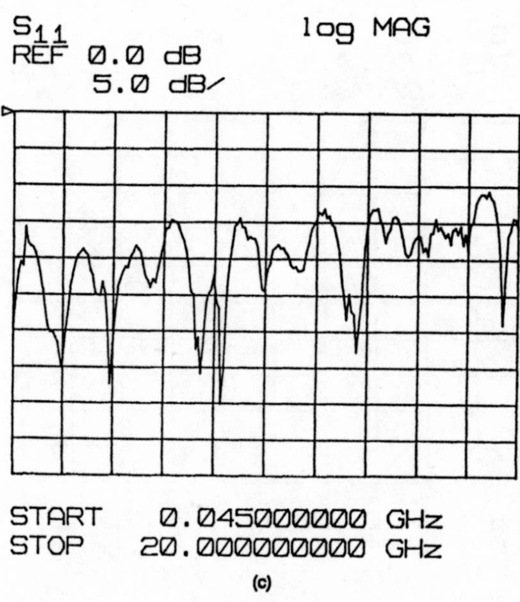

(c)

Figure 4.9 (continued) Performance for the 13.5-mm long −3 dB quadrature coupler on alumina substrate: (b) measured coupled and direct ports, (c) input reflection.

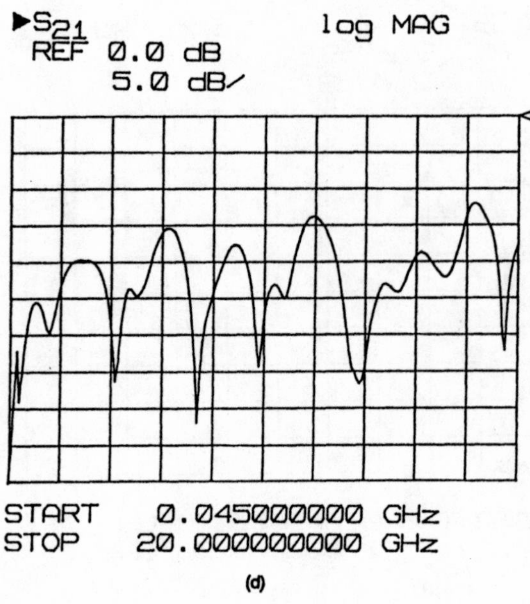

(d)

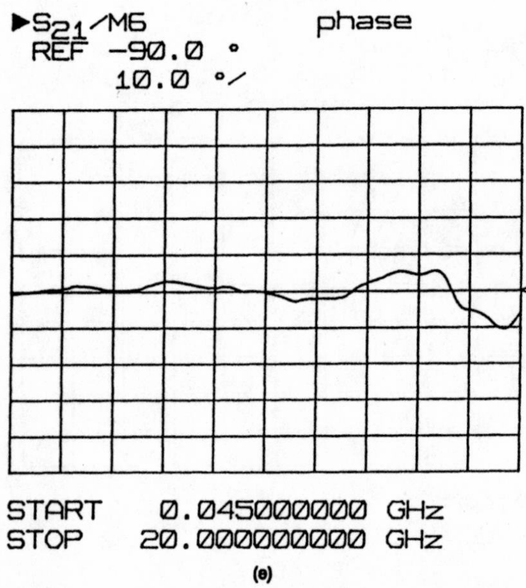

(e)

Figure 4.9 (continued) Performance for the 13.5-mm long −3 dB quadrature coupler on alumina substrate: (d) isolation, (e) phase quadrature.

We can deduce from the measured results that specifying a constant phase velocity increases ripple amplitude, as we predicted in design step 37. However, instead of evaluating cubic splines for the compensated phase velocities, we kept the average value to find its effect on the measured coupler performance. This causes an increase of about 0.4 dB on the ripple amplitude. Therefore, with a maximally flat coupler design, we can expect a ± 0.4 dB ripple when the average phase velocity is used.

4.4.4 1–6 GHz, −3 dB Coupler on Alumina Substrate with $\epsilon_r = 9.9$ and $h = 0.635$ mm

Set the initial specifications: $Z_0 = 50\ \Omega$, $f_c = 3.5$ GHz, $N = 2$, $|C(\omega)|_s = 0.707$, $B_w =$ 1–6 GHz, $r = \pm 0.1$ dB. Choose the phase velocity, an initial coupler length, and step sizes for ω and l: $v = 108 \times 10^9$ mm/s, $l = 5l_c = 38.57$ mm, $\Delta l = 0.05$ mm, and $\Delta\omega = 0.1$.

We have seen in the previous example that we need tandem connection of two −8.34-dB couplers to achieve −3 dB coupling. Therefore, we go directly to design step 30 and choose phase velocities for the interdigitated center section:

$$v_e = 116 \times 10^9 \text{ mm/s} \qquad v_o = 128 \times 10^9 \text{ mm/s}$$

After the first iteration, we change v_o to 127×10^9 mm/s. In this case the value of q is computed as 0.8, which makes the total coupler length 39.2 mm.

The computed results are given in Figure 4.10. It is interesting to note the changes in the computed $p(x)$, which are caused by selective optimization of ripple amplitudes. The new input coupling function suggested in design step 26 gives a performance with equal ripple amplitude. We have seen in the previous example that 10:1 bandwidth is achievable with a five-section design. Therefore, plenty of room is available for further optimization to achieve the theoretical target of -3 ± 0.1 dB. On the other hand, we could have incorporated $v(k)$ using inverse cubic spline interpolation, which would then make the synthesis almost exact (depends on how accurate the phase velocity is predicted). However, this requires more iterations for the optimization. Therefore, whenever possible, the simplest approach should be used. And, of course, we shall now see whether this approach is always correct (satisfactory) or not.

The value of $k(0)$ in this case is computed 0.611, which requires $s_c = 0.101$ mm and $w_c = 0.08$ mm. The function $k(x)$ is shown in Figure 4.10(b). The continuous physical dimensions obtained by evaluating the cubic splines for $w(k)$, $s(k)$, and $d(k)$ for the wiggly sections and $w(k)$ and $s(k)$ for the interdigitated sections at all values of $k(x)$ are shown in Figure 4.10(c).

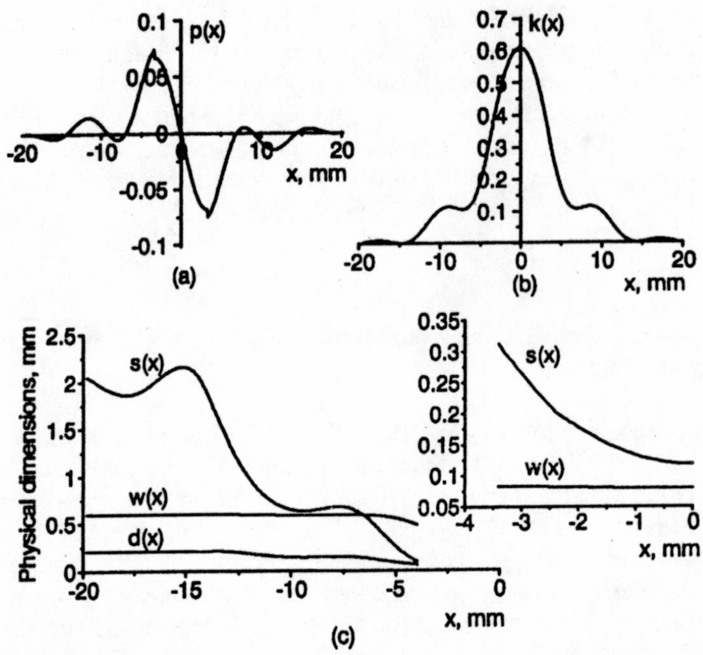

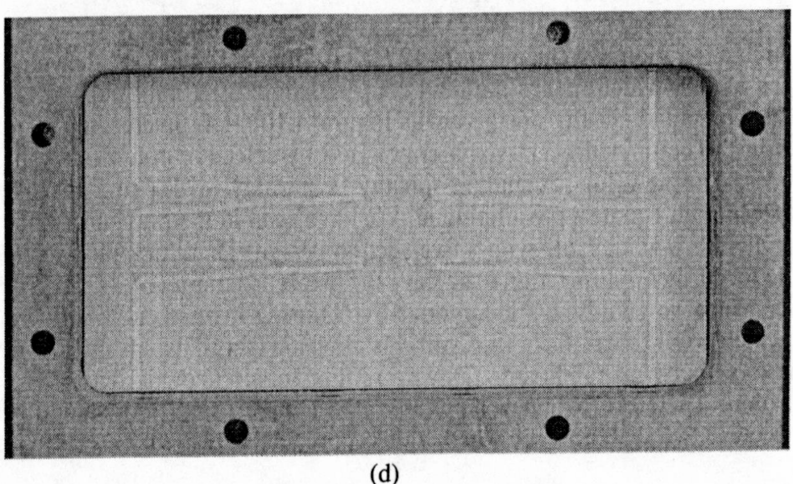

(d)

Figure 4.10 Computed results for the 1–6 GHz, −3 dB coupler on 0.635-mm thick alumina substrate: (a) reflection coefficient distribution function, (b) continuous coupling coefficient, and (c) continuous physical dimensions. (d) The 1–6 GHz, −3dB coupler.

The circuit is realized on a 50.8 × 25.4 × 0.635 mm^3 alumina substrate. Its photograph is shown in Figure 4.10(d). In this design we used optimum mitered bends for the 50 Ω lines. The 50 Ω connecting lines for tandem configuration are 3 mm long.

The computed coupling balance is given in Figure 4.11(a). It is optimized to give a ±0.1-dB ripple in the design bandwidth. The measured one is shown in Figure 4.11(b). The coupling imbalance is shown in Figure 4.11(c). At 6 GHz the imbalance is 2 dB, giving a deviation of 1 dB from the target value. These results show that the

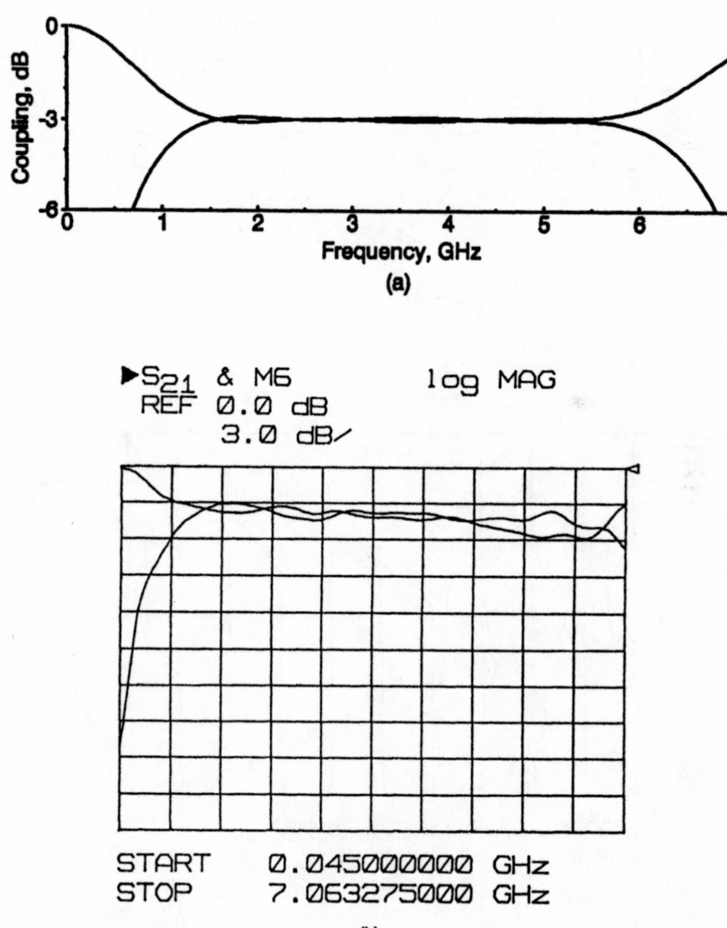

Figure 4.11 Performance for the 1–6 GHz, −3 dB coupler: (a) computed coupling balance, (b) measured coupling balance.

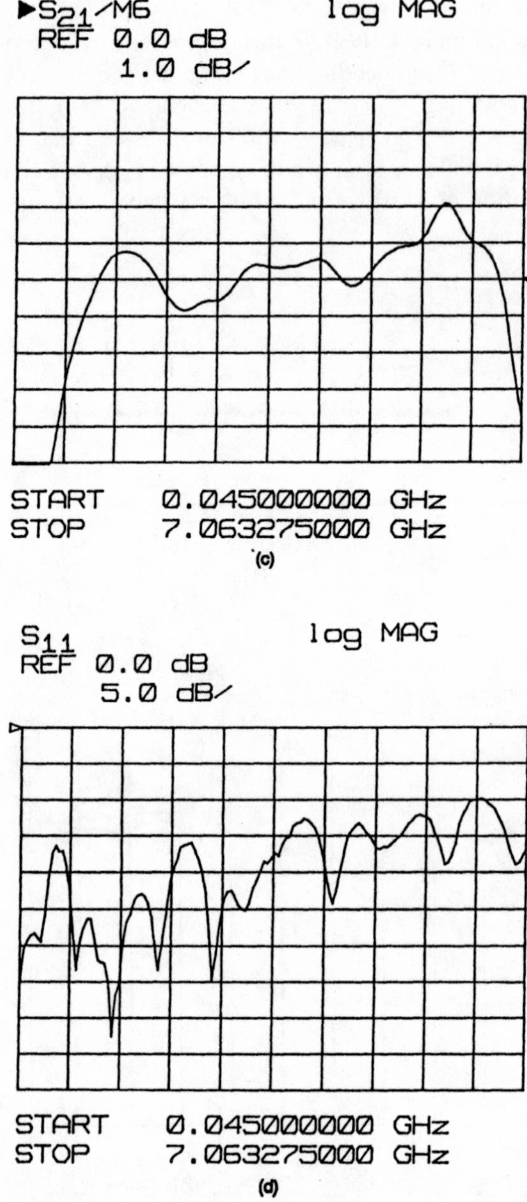

Figure 4.11 (continued) Performance for the 1–6 GHz, −3 dB coupler: (c) measured coupling imbalance. (d) input reflection.

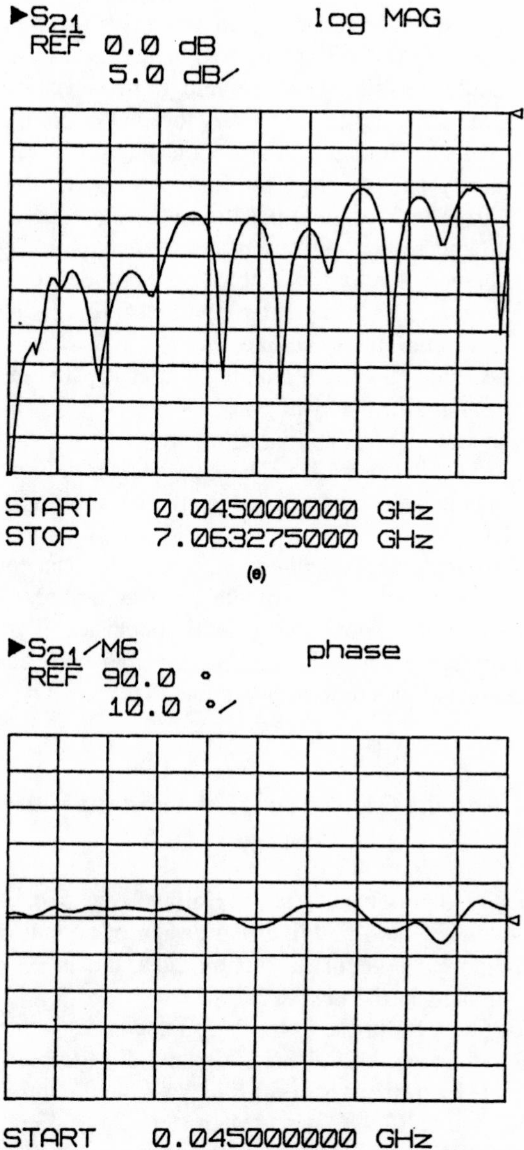

Figure 4.11 (continued) Performance for the 1–6 GHz, −3 dB coupler: (e) isolation, and (f) phase quadrature.

initial prediction is generally correct: using an average phase velocity increases the ripple amplitude by about ± 0.4 dB. At this point, let us take a closer look at the computed $k(x)$, $s(x)$ (double coupled) and the photograph of the design. About 8 mm of each wiggly coupled line setion is seen to be very loosely coupled. In these regions the compensated phase velocity remains almost constant. This causes about an 8% increase in the frequency bandwidth. Therefore, the ripples that were minimized increasingly deviates from their original positions causing an accumulation of excess coupling at the upper band edge as a result of reduced coupler length. In conclusion, we can say that an average value for the compensated phase velocity can generally be used if the desired ripple amplitude is greater than ± 0.4 dB, provided that the ratio of weakly coupled regions to the tightly coupled regions is 1. When this ratio is greater than 1, a new average value should be computed after the first iteration. Otherwise, $v(k)$ should be incorporated in the synthesis.

The measured input reflection and isolation are around -15 dB, except that the latter rises to about -11.5 dB at 5 GHz. These are shown in Figure 4.11(d) and (e). Phase quadrature is maintained within $\pm 3.5°$ in the entire design bandwidth and is given in Figure 4.11(f).

So far, the design of ultrawideband nonuniform directional couplers is discussed as on alumina substrate, which is one of the most commonly used dielectrics in microstrip integrated circuit design. The general conclusion is that, as long as the manufacturing tolerances are kept within ± 5 μm, high performance circuits with every competitive characteristics (compact, simple, planar) can be designed in excess of 10:1 bandwidth.

4.4.5 3–22.5 GHz, -10 dB Coupler on GaAs Substrate with $\epsilon_r = 12.9$ and $h = 0.2$ mm

Another technologically important substrate is Gallium Arsenide, in which real circuit intergration is possible. We shall highlight the design of nonuniform ultrawideband coupler design with a loose coupler in this section and a tandem -3 dB coupler (with a slightly different approach) in the next section.

The initial specifications for the -10 dB coupler are $Z_0 = 50\ \Omega$, $f_c = 12$ GHz, $N = 2$, $|C(\omega)|_s = 0.316$, $B_w = 3$–22.5 GHz, $r = \pm 1$ dB. Choose the phase velocity, initial coupler length, and step sizes for ω and l: $v = 92 \times 10^9$ mm/s, $l = 3l_c = 5.7$ mm, $\Delta l = 0.05$ mm, $\Delta\omega = 0.1$. Then follow design steps 24 to 33.

The initial computation shows that the coupling function is realizable using double-coupled lines only. The optimized computed results are given in Figure 4.12. The maximum value of $k(x)$, Figure 4.12(b), is 0.49, which requires a conductor separation of 0.0161 mm and a width 0.0941 mm on a 0.2-mm thick GaAs substrate. The dimensions at $x = -2.85$ mm are $w = 0.1413$ mm and $s = 0.4478$ mm.

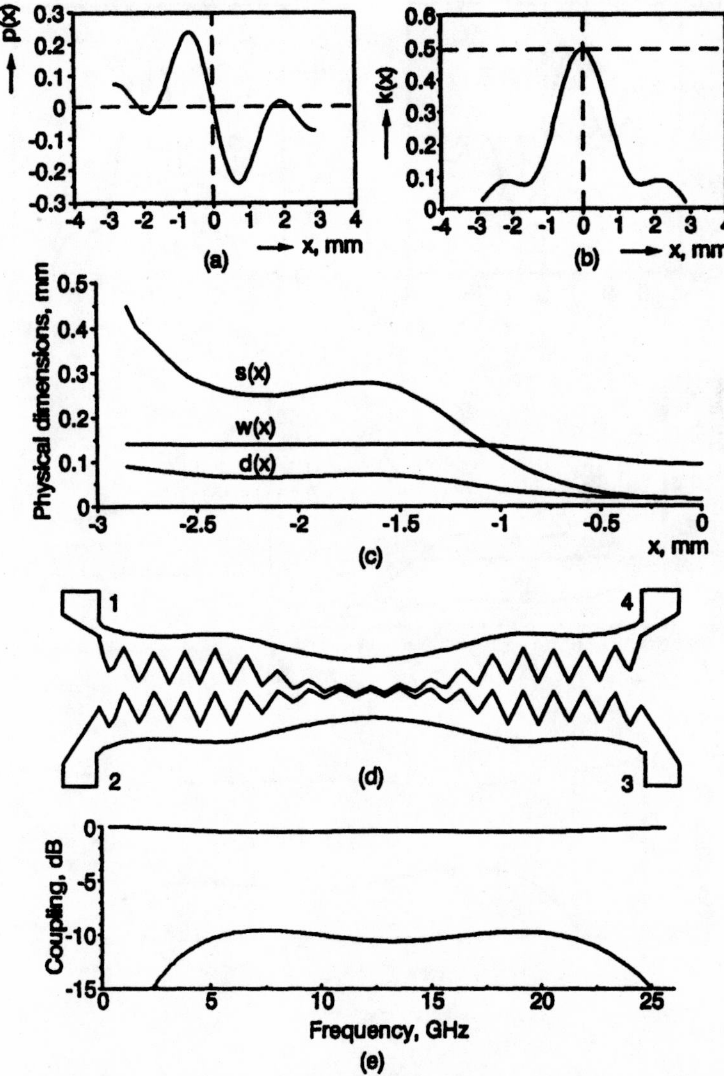

Figure 4.12 Computed results for the 3–22.5 GHz, −10 dB coupler on 0.2-mm-thick GaAs substrate: (a) reflection coefficient distribution function, (b) continuous-coupling coefficient, (c) continuous physical dimensions, (d) coupler layout, and (e) coupled and direct signals.

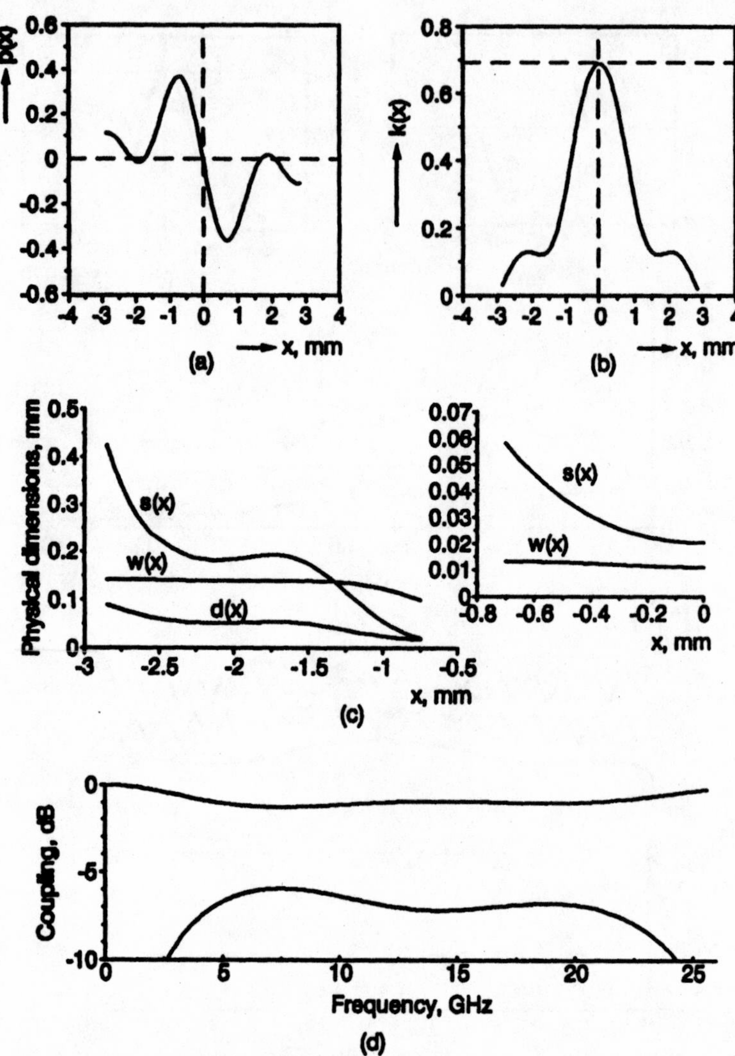

Figure 4.13 Computed results for the 3–22.5 GHz, −3 dB coupler on 0.2-mm thick GaAs substrate: (a) reflection coefficient distribution function, (b) continuous coupling coefficient, (c) continuous physical dimensions, and (d) coupled and direct signals.

Therefore, excluding the connecting 50 Ω lines, this coupler occupies a rectangular area of $5.7 \times 1.037\ mm^2$, which can be fabricated on GaAs.

The physical dimensions $s(x)$, $w(x)$, and $d(x)$ are given in Figure 4.12(c). The computed wiggle depth corresponds to $\Delta x = 0.1$ mm. With this wiggle length the number of subdivisions in each wiggle can be set to 2 ($m = 2$). As long as the resultant wiggle depth is no more than $w(x)/2$, Δx can be changed as desired because very fine tolerance control is possible with GaAs technology. The layout for this design is illustrated in Figure 4.12(d); its computed performance is given in Figure 4.12(e).

4.4.6 3–22.5 GHz, −3 dB Coupler on GaAs Substrate with $\epsilon_r = 12.9$ and $h = 0.2$ mm

First, set the initial specifications: $Z_0 = 50\ \Omega$, $f_c = 12$ GHz, $N = 2$, $|C(\omega)|_s = 0.447$, $B_w = 3$–22.5 GHz, $r = \pm 1$ dB. Choose the phase velocity, an initial coupler length, and the step sizes for ω and l: $v = 92 \times 10^9$ mm/s, $l = 3l_c = 5.7$ mm, $\Delta l = 0.05$ mm, $\Delta\omega = 0.1$. Then follow design steps 24 to 33.

The design procedure reveals that a 1.5-mm long interdigitated section is required at the center. The maximum coupling coefficient in this case is found to be 0.685, which requires $w_c = 0.012$ mm and $s_c = 0.020$ mm with $N = 4$. At $x = -0.75$, the conductor width and spacing for the double-coupled section are 0.099 mm and 0.0198 mm, respectively. These dimensions for the interdigitated section at the same point are 0.0132 mm and 0.058 mm, respectively. Therefore, at the transition regions from $N = 2$ to $N = 4$, we have 9 μm difference. So we can reduce the spacing for $N = 4$ at $x = -0.75$ mm by 3 μm to ensure a smooth transition. This modification has negligible effect on the performance of the coupler. The connections for the alternate fingers can be achieved either by underpasses or air bridges. The computed results for this design are given in Figure 4.13.

The nominal coupling for this design is −7 dB as shown in Figure 4.13(d). For −3 dB coupling, this coupler and the −10 dB coupler of the previous section can be connected in tandem. Although two different coupler designs are required, significant improvement in performance can be achieved because one interdigitated section is eliminated. The total chip area for this design is $5.98 \times 2.58\ mm^2$ on 0.2 mm thick GaAs substrate.

REFERENCE

[1] Uysal, S., and A. H. Aghvami, "Synthesis, Design, and Construction of Ultra-Wide-Band Nonuniform Quadrature Directional Couplers in Inhomogeneous Media," *IEEE Trans. on Microwave Theory and Tech.*, Vol. MTT-37, 1989, pp. 969–976.

Chapter 5
Sensitivity Analysis for Nonuniform Line Directional Couplers

5.1 THE EFFECT OF NONUNIFORM SLOPES OF COUPLED LINES

High-performance circuits require a careful evaluation of a number of extrinsic and intrinsic factors that affect the result. Most of these factors may not show up until implementation. Even an experienced design engineer may fail to predict the effect of certain parameters. It is extremely difficult to present a complete mathematical formulation similar to the one for uniform coupled lines [1] for the sensitivity of performance with variation of certain parameters. Among such parameters is the nonuniform slope (taper rate) of coupled lines.

When designing nonuniform couplers, one should roughly know the spacing required with certain coupling coefficient values. For example, 2 mm spacing is required for -30 dB coupling or 0.06 mm for -8 dB coupling (for exact values, refer to Chapter 3). This saves design time and also forces the designer to introduce alternative solutions at the initial design stages.

For coupling the electromagnetic energy from an excited line to an adjacent line we assume a perpendicular interaction of fields. Under this assumption, we have presented a transformation technique from uniform to nonuniform parameters and vice versa. The validity of this assumption decreases with very sharp taper rates. This situation usually arises at the coupler edges with relatively short coupler lengths, which is illustrated in Figure 5.1(a). In this case, everything is done according to the theoretical values obtained from the synthesis. If the computations give an initial coupling value of $k = 0.00016598$ the corresponding dimensions are then obtained as $s = 8$ mm, $w = 0.625$ mm for alumina substrate with $\epsilon_r = 9.9$ and $h = 0.635$ mm. This approach is wrong. One may assume that a certain value of spacing, for example,

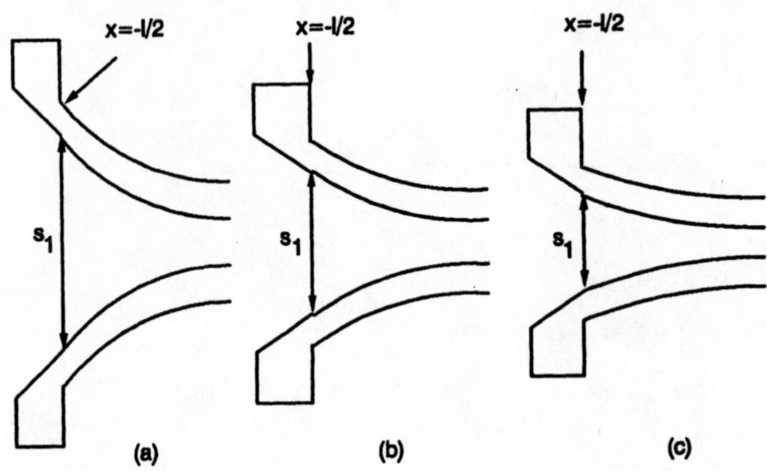

Figure 5.1 Selection of initial spacing: (a) minimum spacing is computed from theoretical (!) value, (b) best guess, and (c) spacing is set to a value that can be safely predicted with the analysis.

seems to be very accurate or exact, extreme ends should be treated cautiously. In the case of coupled lines, surface waves that we usually neglect may cause around -40 dB spurious coupling.

The solution to this problem was given in Section 4.4. It is always best to set the initial coupling to a small finite value that can be safely predicted with the analysis. This technique ensures minimum distortion on coupling and therefore on phase quadrature, which may be very critical for some applications by maintaining proper coupling of fields from one line to the other.

The effect of nonuniform slopes becomes more complex when wiggling is introduced into the coupled regions. It can become so complex that we may decide to abandon the design. This situation is illustrated in Figure 5.2. We have two extreme cases of wiggly coupled lines. For the first case, we use weakly coupled lines, as shown in Figure 5.2(a). Spacing is taken as the vertical distance from any point on the inner edge of the first line to the inner edge of the second line. In the second extreme we have an almost complete overlap of wiggles, as illustrated in Figure 5.2(b). In this situation s cannot be taken as the vertical distance between the two inner edges. Because coupling is defined as the ratio of the difference between the even- and odd-mode impedances to their sum at any point along the coupler, we can safely explain this situation. The even-mode impedance is almost unaffected because the effect of wiggling on the even-mode capacitance parameter C_{fe} is almost negligible. However, for the odd mode the capacitance parameter C_{fo} significantly increases with decreasing spacing. Therefore,

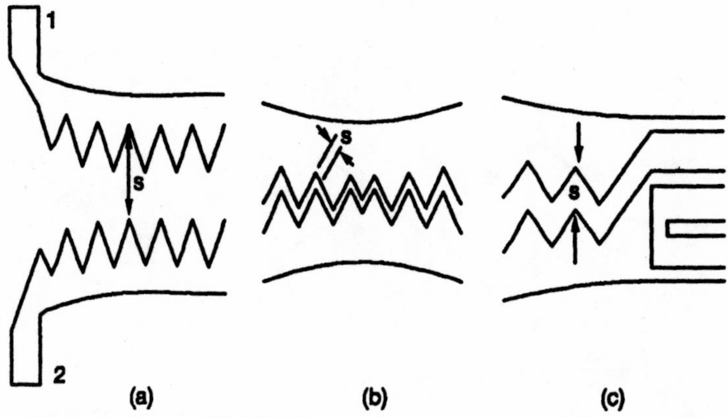

Figure 5.2 Illustration of a design gap with wiggly coupled lines: (a) with a large gap, (b) a very narrow gap, and (c) adjusting the center section length to avoid an extreme overlap of wiggles.

the dominant coupling parameter is C_{fo}, which travels the additional path introduced by wiggling. It is usually the case that with wiggly uniform couplers a slightly higher coupling is reported for a given s compared to couplers with no wiggle. In the case of nonuniform couplers this situation can be avoided by taking the necessary measures. For tight (-3 dB tandem) couplers, because we introduce an interdigitated center section, the variable q can be adjusted to change the length of the interdigitated section so that extreme overlapping of wiggles is avoided or at least this region is kept fractionally small compared to a quarter-wavelength at the design center frequency. This is shown in Figure 5.2(c). Alternatively, this procedure may be followed: when the wiggles begin to overlap, calculate the new vertical spacing with the help of Figure 5.3; the new value of spacing denoted by s_{new} is calculated so that the value of s becomes the design spacing at that point, or adjust coupling coefficient function $k(x)$ at these regions such that the new $k(x)$ gives the desired spacing.

This extreme situation usually arises when the coupler length is long (design example 4.4.4). It has been my experience that the fastest way of getting around such problems is to adjust the value of q. In the case of loose couplers that require no interdigitated section, it is best to use Figure 5.3.

One final reminder. The continuously varying slope results in a slightly longer coupler length, which may cause a slight shift of the design center frequency. This additional length can be calculated by computing the slopes given in Chapter 4 and then computing the approximate inclined elemental lengths. It is usually between 1 and 3% depending on the axial length of the coupler.

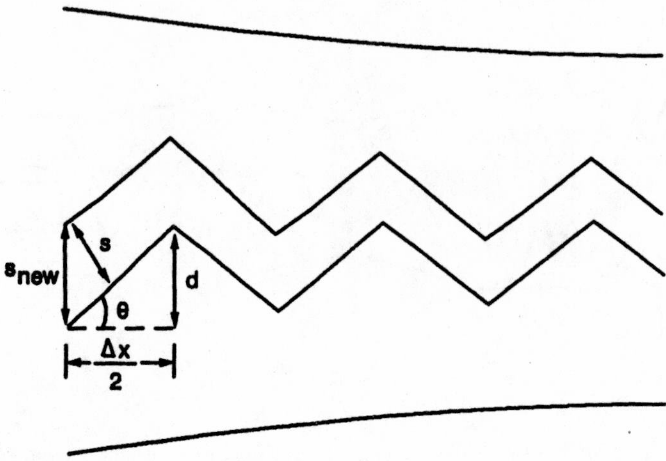

Figure 5.3 We need to compute the value of s_{new} with overlapping wiggles so that the formulation for the geometry (Chapter 4) can be used with no additional modification.

5.2 THE EFFECT OF FABRICATION TOLERANCES

There is an increasing trend in manufacturing MICs to use plate-up (*e*-beam lithography) or dry etch processes. The aim is to eliminate manufacturing tolerances in order to realize high-grade performance from circuits demanding a tight control on dimensional variations. This can easily be justified from the ever-increasing communications traffic. The available bandwidth is limited (one has to define an operational bandwidth, which automatically imposes a restriction on it), and the user may not like interference from other channels. This situation can be directly translated to the method of production of certain components at the laboratory. However, the effect of manufacturing tolerances is based on the assumption that theoretical prediction is the nearest to that of an ideal (!) circuit performance. With this argument in mind, we still want to find out the results of our design, which may be based on certain assumptions. This clearly means that maintaining the exact design dimensions with the fabrication technique may not always result in working circuits. Now we can proceed to identify those physical dimensions in nonuniform directional couplers that are most likely to affect the overall performance.

The etching solutions are prepared from highly reactive chemicals. The required percentages for the etchants are usually supplied by the manufacturers along with recommended etching times and temperatures. There is no need to say "follow these guidelines." Once the amount of overetching (this is usually the case, as underetching is related to dimensional errors in mask production and photoresist developing stages) is determined, the dimensions may be adjusted accordingly. However, in nonuniform

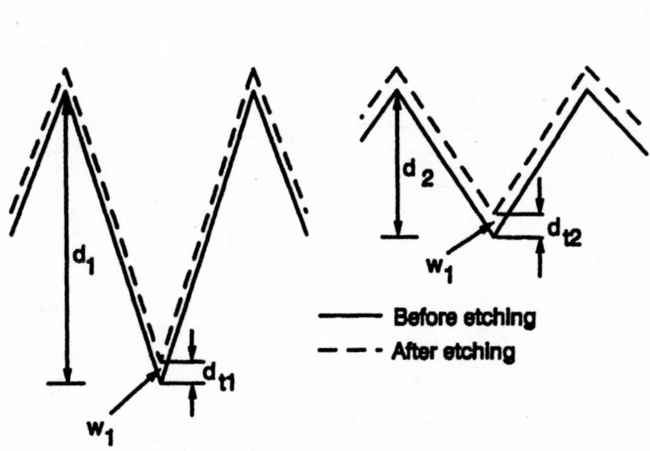

Figure 5.4 Etching tolerances on wiggle depths.

directional couplers we employ continuously varying wiggle depths. Therefore, for each wiggle we need to define a different tolerance. This is illustrated in Figure 5.4.

For a given fixed tolerance, the effect on d_1 will be higher. It is clear from this figure that along with etching tolerances we also need to define the minimum realizable linewidth (w_t) and spacing (s_t) (in fact, these should be considered at the masking stage). Consider a numerical example. Set the etching tolerance as 15 μm. This implies that $w_1 = 2 \times 15$ μm $= 30$ μm. With $\Delta x = 0.3\ d_1 = 0.6$ mm, and $d_2 = 0.15$ mm, the values for d_{t_1} and d_{t_2}, as shown in Figure 5.4, can be calculated as 0.06 mm and 0.03 mm, respectively. This shows that it is not possible simply to add a constant value to the computed wiggle depth to compensate for a predicted amount of overetching. The new wiggle depth with tolerances should be computed for each wiggle. If you are designing a wiggly nonuniform coupler for the very first time, it is highly recommended to stick with the computed values. This is because etching tolerances and definitions may vary considerably for different techniques and different persons doing the etching. In this way, we can identify the exact nature of possible overetching.

5.3 THE EFFECT OF BOND WIRES AND TRANSITION FROM DOUBLE COUPLED LINES TO AN INTERDIGITATED CENTER SECTION

We have seen in the previous chapter that, for ultrawideband nominal coupling between -14 dB and -7 dB, we need to use an interdigitated section at the coupler

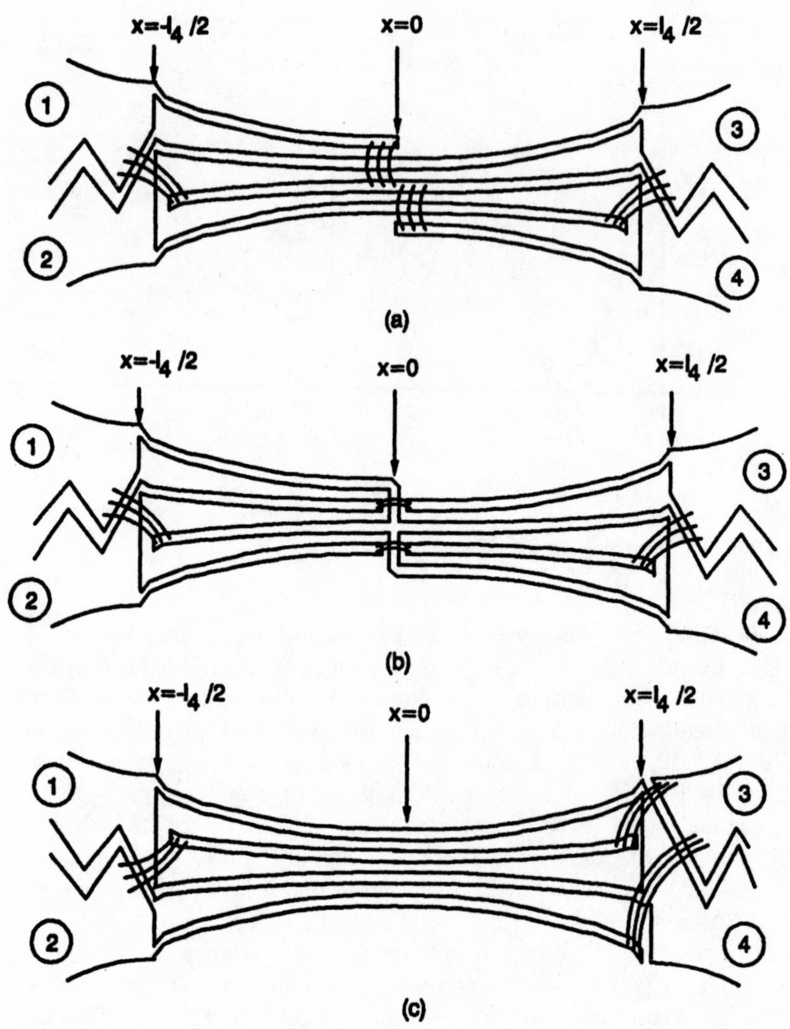

Figure 5.5 Interdigitated center section configurations: (a) original Lange type, (b) modified connections at $x = 0$, (c) asymmetrical with fewer bond wires.

center. Depending on the application a slightly different interdigitated section may be used. Some possible configurations are shown in Figure 5.5. The original configuration for the Lange coupler [2] can be used in nonuniform form; this is shown in Figure 5.5(a). Bond wires [3] are required at four different locations to connect alternate fingers. An alternative configuration obtained by modifying the region at $x = 0$ is given in Figure 5.5(b). In this configuration the bond wires in the vicinity of $x = 0$ are

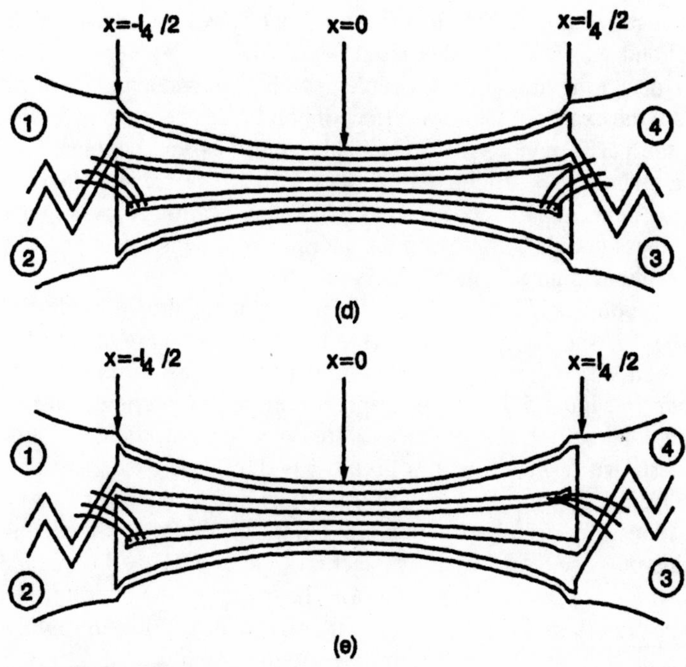

Figure 5.5 (continued) Interdigitated center section configurations: (d) unfolded Lange type, (e) unfolded symmetrical.

parallel to the fingers. This arrangement of bond wires is useful when spacing at $x = 0$ is large and f_c is high. Both of these configurations provide a balanced connection for direct and coupled signals. These arrangements also give the two outputs on the same side of the coupler, which is a practical advantage in tandem connection and some applications. Another possible configuration, which again gives outputs on the same side of the coupler, is shown in Figure 5.5(c). For this one we use less bond wires but we have asymmetrical connections at the transitions at $x = -l_4/2$ and $x = l_4/2$.

When there is no requirement for adjacent outputs, we can use the unfolded version of the Lange coupler [4], which is given in Figure 5.5(d). This arrangement is asymmetrical but the layout is straightforward. Bond wires are required at only two locations. An alternative configuration giving electrically symmetrical transitions at $x = -l_4/2$ and $x = l_4/2$ is shown in Figure 5.5(e). In this case the layout is asymmetrical and again we require bond wires only at two locations.

In Figure 5.5, we have shown five possible different configurations for the center section. Depending on the application anyone of them can be selected. It can be seen from these configurations that we have bond wires with different orientations. In mass production, it may be desirable to have the bond wires parallel to each other. The location of bond wires will be discussed next.

When designing directional couplers with ultrawideband performance, it is difficult to see the exact nature of the effect of bond wires because several other factors may affect the performance. However, with bandpass-type couplers we can design moderate bandwidth -3 dB couplers with or without an interdigitated section, as we shall see in the next chapter. The effect of bond wire location is investigated by using a -3 dB bandpass coupler employing an interdigitated center section. The bond wire locations for this design are shown in Figure 5.6.

First, the coupler is bonded using the arrangement shown by dashed lines. The measured result is shown in Figure 5.7(b). The bond wires are then removed and the coupler is rebonded as shown by solid bond wires. The measured performance in this case is shown in Figure 5.7(c), which agrees well with the computed result given in Figure 5.7(a). Therefore, irrespective of orientational difficulties in bonding, the arrangement shown in solid lines in Figure 5.6 should be used.

The physical length of a nonuniform directional coupler can be related to its coupling response by dividing the coupler length into quarter-wavelength-long sections. For this purpose, consider a five-section ($l = 5\ \lambda c/4$) -3 dB tandem coupler, one-half of which is given in Figure 5.8(a). The length of the interdigitated section is assumed to be $\lambda c/4$ ($q = 1$) at the design center frequency. When designing directional couplers we always specify an allowable ripple amplitude as $\pm r$ in the design bandwidth (with maximally flat coupling $r = 0$).

The transition points (from $N = 2$ to $N = 4$ and vice versa) correspond to -3 dB coupling points as indicated on Figure 5.8(b). It is very important to maintain a smooth transition at these regions; otherwise, they may cause unacceptable mismatches. However, when $k(x)$ is transformed into physical dimensions we usually end up with two possible situations, which are shown in Figure 5.9.

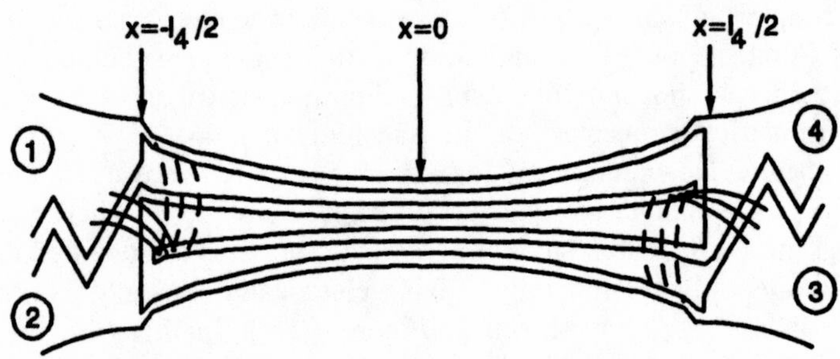

Figure 5.6 Center section for the investigation of the effect of bond wire location. Dashed lines show parallel bond wires perpendicular to the fingers. Solid bond wires use the shortest distance from the first wiggle.

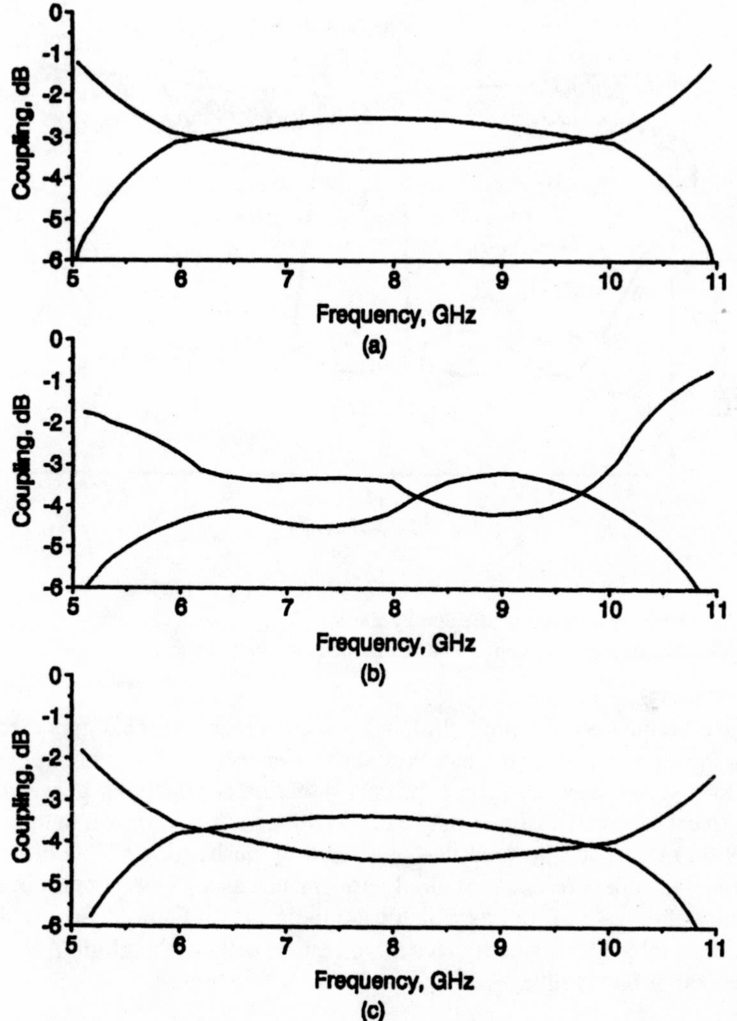

Figure 5.7 Effect of bond wire location on the performance: (a) computed results, (b) measured results with dashed-line bond wires shown in Figure 5.6(a), and (c) measured results with solid-line bond wires. In both cases a maximum of three bond wires were used for each location.

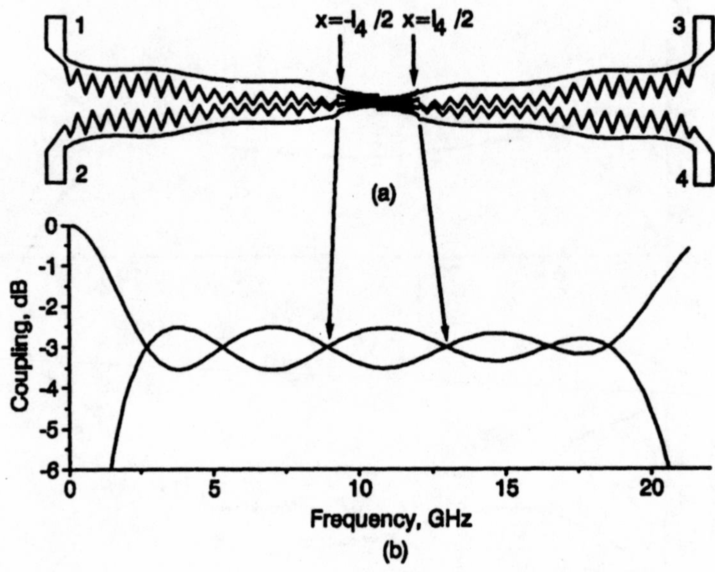

Figure 5.8 Frequency regions most affected by the transition regions: (a) -8.34 dB coupler with interdigitated center section and (b) its tandem performance.

In the extreme case, Figure 5.9(a), a redesign is required. This may mean a new value for q in order to change the length of the center section. We cannot possibly use the previously synthesized $k(x)$ since the center section has different phase velocities. However, an experienced designer may end up with a transition shown in figure 5.9(b). The transition region can be modified as shown by dashed lines. The effect of this modification will reflect on the measured performance as a slight increase in coupling at the vicinity of the -3 dB points which were indicated on Figure 5.8(b). This is well within the acceptable limits since such a modification will have negligible effect on the rest of the design bandwidth.

5.4 THE EFFECT OF Δl

It is possible to choose a very small value for Δl in the computations, which undoubtedly will give the best results. However, this is not cost effective and may also cause some memory problems for the storage of the computed data values; also dealing with a large number of data becomes more difficult at the mask generation stage. This is also true with long couplers.

The effect of Δl is investigated with a nonoptimum five-section -8.34 dB coupler. The computed $k(x)$ for different Δl values in the vicinity of the coupler center

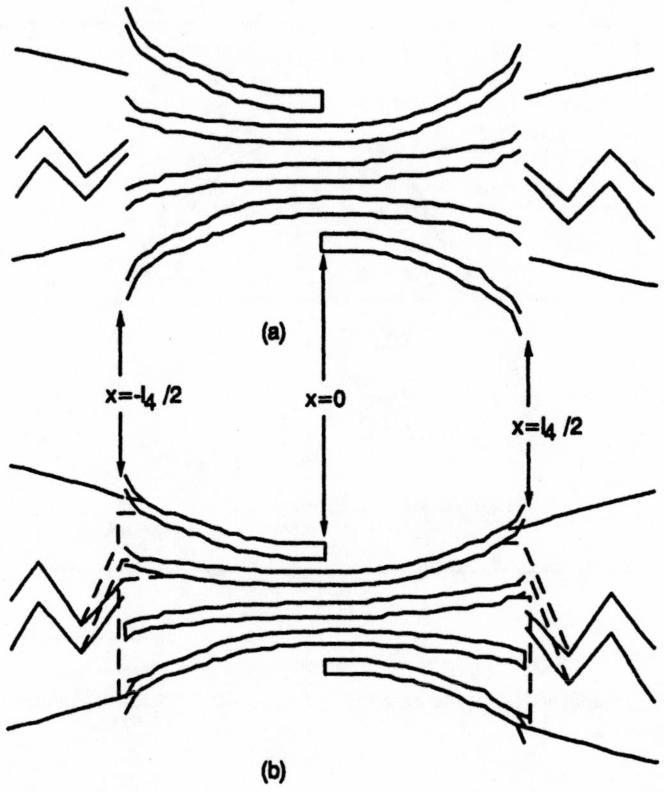

Figure 5.9 Transitions from double-coupled lines to an interdigitated center section: (a) extreme case with $3s_4 + 4w_4 \gg 2w_2 + s_2$ and (b) modification of transition regions to ensure a smooth transition from $N = 2$ to $N = 4$.

are given in Figure 5.10. There $k(x)$ at $x = 0$ sharply converges for $\Delta l < 0.2$ mm. The corresponding $k(0)$ for $\Delta l = 0.2$ mm is 0.702 whereas this value for $\Delta l = 0.005$ mm is 0.7225, a mere 2.9% change in $k(0)$ for a 40-fold reduction in Δl. Therefore, it is usually sufficient to choose an easily manageable value for Δl.

5.5 THE EFFECT OF AN INTERDIGITATED SECTION

Employing wiggling for the inner edges of double-coupled lines effectively reduces isolation in an ultrawideband design bandwidth. However, the interdigitated center section has different phase velocities that affect the overall performance. That is, the scattering parameters become functions of these phase velocities in this section.

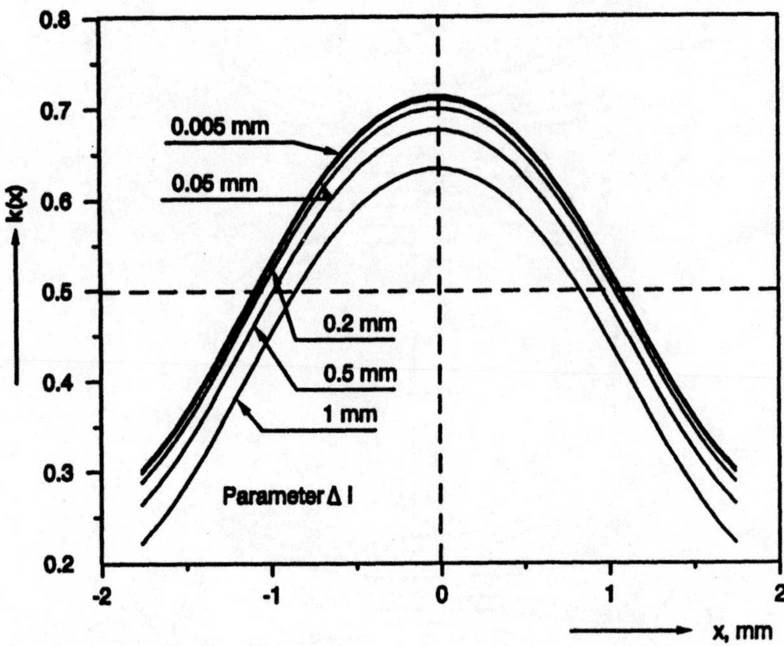

Figure 5.10 The computed $k(x)$ at the center of a five-section -8.34 dB coupler with different elemental lengths.

Detailed formulation for this will be given in Chapter 9. For the time being, let us assume that the effect of phase velocities on S_{11} and S_{31} is the same. A mismatch at the input means less coupled signal. With the same reasoning, we can say that a finite S_{31} means less direct signal. This oversimplified explanation is very helpful because it provides immediate information on the quality of the measured performance.

Consider a lossless directional coupler:

$$|S_{11}|^2 + |S_{21}|^2 + |S_{31}|^2 + |S_{41}|^2 = 1 \tag{5.1}$$

We propose

$$|S_{11}|^2 + |S_{21}|^2 = 0.5 \tag{5.2}$$

$$|S_{31}|^2 + |S_{41}|^2 = 0.5 \tag{5.3}$$

In a practical coupler, we have

$$|S_{11}|^2 + |S_{21}|^2 = 0.5 - \text{Loss} \tag{5.4}$$

$$|S_{31}|^2 + |S_{41}|^2 = 0.5 - \text{Loss} \tag{5.5}$$

These relationships help us identify possible overetching and design errors. As an exercise, apply these relationships to the measured results given in Chapter 4.

Finally, the following relationship can be used to determine the effect of unequal phase velocities on the phase quadrature:

$$\Delta\theta = 90 \pm \tan^{-1}\left[\frac{2S_{21}\sqrt{1 - S_{21}^2}}{S_{31}}\right] \tag{5.6}$$

where $\Delta\theta$ is the deviation from 90°, the plus sign is for phase (S_{21}), and the minus sign is for $(S_{41}) > \pi/2$.

5.6 THE EFFECT OF LOSSES

The effect of ohmic and dielectric losses on the performance of uniform microstrip directional couplers has been reported by Rao [5]. In the case of nonuniform directional couplers it may seem easy enough to introduce the loss factor α in the propagation constant γ in the derivation of the nonlinear differential equation described in Chapter 1. However, its determination becomes extremely complex. In addition to ohmic and dielectric losses, radiation loss plays an important role on the overall performance. When designing −3 dB couplers, we can assume that coupled and direct signals have roughly the same amount of loss that does not affect the coupling balance. If the losses are unacceptable, alternative designs in the form of shielding and suspending may be considered. For other couplings, a knowledge of the loss tangent of the chosen substrate and surface roughness of the conductor are useful in order to have an idea of ohmic and dielectric losses. Alternatively, we can design a coupler and use it as a reference for other couplers.

REFERENCES

[1] Shamasundara, S. D., and K. C. Gupta, "Sensitivity Analysis of Coupled Microstrip Directional Couplers," *IEEE Trans. on Microwave Theory and Tech.*, Vol. MTT-26, 1978, pp. 788–794.

[2] Lange, J., "Interdigitated Stripline Quadrature Hybrid," *IEEE Trans. on Microwave Theory and Tech.*, Vol. MTT-17, 1969, pp. 1150–1151.

[3] March, S. L., "Simple Equations Characterize Bond Wires," *Microwaves & RF*, November 1991, pp. 105–110.

[4] Waugh, R., and D. LaCombe, "Unfolding the Lange Coupler," *IEEE Trans. on Microwave Theory and Tech.*, Vol. MTT-20, 1972, pp. 777–779.

[5] Rao, B. R., "Effect of Loss and Frequency Dispersion on the Performance of Microstrip Directional Couplers and Coupled Line Filters," *IEEE Trans. on Microwave Theory and Tech.*, Vol. MTT-22, 1974, pp. 747–750.

Chapter 6
Bandpass Nonuniform Line Directional Couplers

6.1 INTRODUCTION

It is possible to achieve tight coupling up to almost 0 dB using nonuniform double-coupled lines without the need for tandem connection, an interdigitated section, or extreme photolithographic techniques. This is made possible by frequency-selective coupling in nonuniform line couplers. The result is very high-performance, entirely planar (no bond wires) directional couplers with directivities in excess of 30 dB because wiggling effectively reduces isolation. This technique requires a suitable modification of the reflection coefficient distribution function associated with nonuniform directional couplers [1].

6.2 DESIGN PROCEDURE

For a matched nonuniform directional coupler, the reflection coefficient distribution function can be expressed as [1,2]

$$p(x) = -\frac{2}{\pi v}\int_0^{2\omega_c} \sin(2\omega x/v)\tanh^{-1}[C(\omega)]\,d\omega \tag{6.1}$$

where v is the compensated phase velocity, ω_c is the design center frequency, and

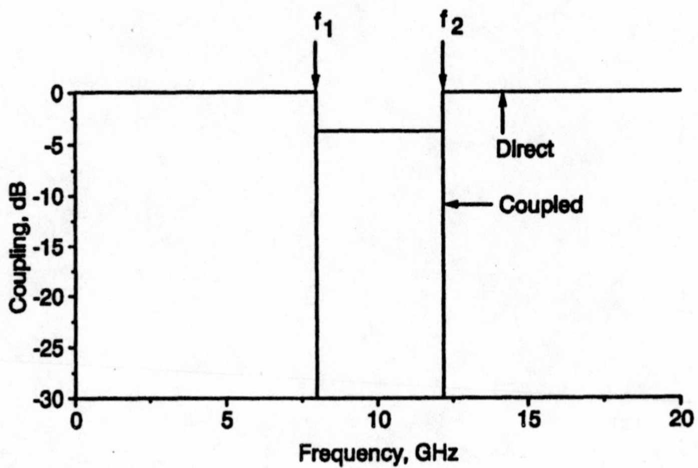

Figure 6.1 Bandpass coupler principle.

Equation (6.1) forms a Fourier transform pair with the coupled-arm response that follows:

$$C(\omega) = \tanh\left\{ \int_0^l \sin(2\omega x/v) p(x) dx \right\} \tag{6.2}$$

where l is the coupler length.

It can be shown by using simple integration theorems that, if a function is continuous except for a finite number of finite discontinuities in the interval of integration, the function is integrable. In Figure 6.1 we have two discontinuities at the band edges f_1 and f_2. Therefore, for a bandpass coupler, the reflection coefficient distribution function can be modified as follows:

$$p(x) = -\frac{2}{\pi v}\left\{ \int_0^{\omega_1} \sin(2\omega x/v) \tanh^{-1}[C_1(\omega)]\, d\omega + \int_{\omega_1}^{\omega_2} \sin(2\omega x/v) \tanh^{-1}[C_2(\omega)]\, d\omega \right.$$
$$\left. + \int_{\omega_2}^{2\omega_c} \sin(2\omega x/v) \tanh^{-1}[C_3(\omega)]\, d\omega \right\} \tag{6.3}$$

where $\omega_1 = 2\pi f_1$, $\omega_2 = 2\pi f_2$, and $C_1(\omega)$, $C_2(\omega)$, and $C_3(\omega)$ are the specified responses in the respective bands. $C_1(\omega)$ and $C_3(\omega)$ can be set to zero.

The step-by-step design procedure given in Chapter 4 equally applies to bandpass-type couplers with $p(x)$ replaced by eq. (6.3). However, for the bandpass coupler, because we define $\omega_1 > 0.0$, the computed coupling coefficient becomes unrealizable.

To have a realizable continuous coupling coefficient, the following relationship must be satisfied:

$$Z_{0e}(x) \geq 1.0 \qquad \forall x \tag{6.4}$$

where $Z_{0e}(x)$ is the normalized even-mode characteristic impedance of the coupler.

The continuous coupling coefficient can be written in terms of $p(x)$ as follows:

$$k(x) = \frac{e^{4\int_0^x p(x)dx} - 1}{e^{4\int_0^x p(x)dx} + 1} \tag{6.5}$$

where $p(x)$ is given by (6.3). There is a nonlinear relationship between $k(x)$ and $p(x)$. This means that $p(x)$ can be modified in a number of ways to achieve a realizable $k(x)$.

6.3 MODIFICATIONS OF $p(x)$

For a realizable $k(x)$, $p(x)$ may not be a positive function for all negative x values but the initial value of $p(x)$, that is, $p(-l/2)$, must always be positive. The principle of realizability is illustrated in Figure 6.2.

For a quadrature performance, we require $p(x)$ to have odd symmetry. For a realizable $k(x)$ we require

$$\int_{-l/2}^{-x_1} p(x)dx > 0$$

$$\int_{-x_1}^{-x_2} p(x)dx - \int_{-l/2}^{-x_1} p(x)dx > 0$$

$$\int_{-x_2}^{-x_3} p(x)dx - \left[\int_{-l/2}^{-x_1} p(x)dx + \int_{-x_1}^{-x_2} p(x)dx\right] > 0$$

$$\int_{-x_3}^{-x_4} p(x)dx - \left[\int_{-l/2}^{-x_1} p(x)dx + \int_{-x_1}^{-x_2} p(x)dx + \int_{-x_2}^{-x_3} p(x)dx\right] > 0$$

$$\int_{-x_4}^{0} p(x)dx - \left[\int_{-l/2}^{-x_1} p(x)dx + \int_{-x_1}^{-x_2} p(x)dx + \int_{-x_2}^{-x_3} p(x)dx + \int_{-x_3}^{-x_4} p(x)dx\right] > 0$$

If $\int p(x)dx$ in any interval is equal to zero this gives $k(x) = 0$, which is not desirable, as explained in Chapters 4 and 5. It can be deduced from the preceding equations that $p(x)$ may have negative values for an interval of negative x values as long as the overall sum is positive.

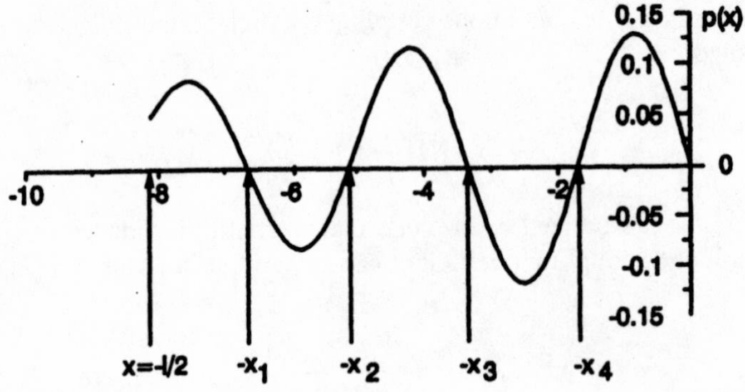

Figure 6.2 Reflection coefficient distribution function for a bandpass coupler. The function has odd symmetry of about $x = 0$.

6.3.1 Introducing a Dummy Channel for Realizability

The simplest way of ensuring a realizable $k(x)$ is to introduce a dummy coupling channel $C_d(\omega)$ in the 0–ω_d frequency range. The new $p(x)$ is obtained by

$$p_d(x) = -\frac{2}{\pi v}\left\{\int_0^{\omega_d} \sin(2\omega x/v)\tanh^{-1}[C_d(\omega)]\,d\omega + \int_{\omega_1}^{\omega_2} \sin(2\omega x/v)\tanh^{-1}[C_2(\omega)]\,d\omega\right\} \tag{6.6}$$

The coupling response is given by

$$C(\omega) = \tanh\left\{\int_{-l/2}^{l/2} \sin(2\omega x/v)p_d(x)dx\right\} \tag{6.7}$$

6.3.2 Adding a Constant to $p(x)$

The reflection coefficient distribution function computed by using (6.3) can be modified by adding a constant [2]:

$$p_A(x) = p(x) + A \tag{6.8}$$

where A is a constant given by

$$A = \begin{cases} a & -l/2 \leq x \leq 0 \\ -a & 0 \leq x \leq l/2 \end{cases}$$

The corresponding coupling function is obtained by

$$C_A(\omega) = C(\omega) - j\frac{Av}{\omega}\sin^2(\omega l/2v) \tag{6.9}$$

The constant A can be determined from the normalized even-mode characteristic impedance function as

$$A \geq \frac{1}{2l}\ln\left(\frac{1}{Z_{0e}(x)_{\min}}\right) \tag{6.10}$$

where $Z_{0e}(x)_{\min}$ is the minimum value obtained with $A = 0$.

6.3.3 Adding a Linear Function to $p(x)$

A linear function may be added on the computed $p(x)$ to achieve realizability:

$$p_B(x) = p(x) - 2\frac{Bx}{l} \tag{6.11}$$

where B is a constant and l is the coupler length. The corresponding coupling function is obtained by

$$C_B(\omega) = C(\omega) - j\frac{Bv}{\omega}\left[\cos(\omega l/v) - \frac{v}{\omega l}\sin(\omega l/v)\right] \tag{6.12}$$

The constant B is given by

$$B \geq \frac{1}{2l}\ln\left(\frac{1}{Z_{0e}(x)_{\min}}\right) \tag{6.13}$$

where $Z_{0e}(x)_{\min}$ is the minimum value obtained with $B = 0$.

6.3.4 Multiplying Even-Mode Impedance by a Constant

We have already seen in Chapter 4 that the initial coupling coefficient can be adjusted by multiplying the normalized even-mode impedance by a constant. In this section, a similar method will be given to modify $p(x)$.

Multiplying the normalized even-mode impedance by a constant, we obtain

$$Z_{0e_E}(x) = Z_{0e}(x)\,e^{E} \tag{6.14}$$

The new reflection coefficient distribution function becomes

$$p_E(x) = p(x) + \frac{1}{2}E\left[\delta\left(x + \frac{l}{2}\right) - \delta\left(x - \frac{l}{2}\right)\right] \tag{6.15}$$

The coupled response is then obtained as

$$C_E(\omega) = C(\omega) + jE\sin(\omega l/v) \tag{6.16}$$

The constant E is given by

$$E > \ln\left(\frac{1}{Z_{0e}(x)_{\min}}\right) \tag{6.17}$$

where $Z_{0e}(x)_{\min}$ is the minimum value obtained with $E = 0$. This way of modifying $p(x)$ is very attractive because it increases the coupling without having to use longer coupler lengths. It also increases the out-of-band coupling, but this is not significant because the maximum coupling can be maintained at a desired value. This modification also causes a shift in the design bandwidth that can be corrected by specifying a new bandwidth.

6.3.5 Selective (Regional) Optimization of $p(x)$

Figure 6.2 showed two regions in which $p(x)$ becomes negative. Because $|p(x)|$ between $-x_3$ and $-x_4$ is only slightly larger than $|p(x)|$ between $-x_2$ and $-x_3$, it may be sufficient to modify the region between $-x_2$ and $-x_3$. A linear, increasing positive function or a positive constant may be added to $p(x)$ in this region. Because mathematical formulation becomes rather complicated, it will not be given. However, a numerical solution is straightforward by substituting the new $p(x)$ into the coupled function. The functions (linear or constant) may be adjusted until a satisfactory $k(x)$ is achieved.

Alternatively, $p(x)$ between $-l/2$ and $-x_1$ may be increased until the realizability conditions are satisfied. Furthermore, $p(x)$ may be modified at any point as long as it gives a realizable $k(x)$. It is clear that selective (regional) optimization can be done in many ways. The aim is to achieve a realizable $k(x)$ and to keep its maximum value within realizable limits with double-coupled lines using ordinary photolithographic techniques. We also must pay attention to the taper rates between coupling coefficient maxima and minima along the coupler length.

6.4 DESIGN EXAMPLES

In this section some of the key designs will be given for realizable bandpass couplers by using a dummy channel or by multiplying the even-mode impedance by a constant value.

6.4.1 K_u-band, -3 dB Bandpass Coupler on Alumina Substrate with $\epsilon_r = 9.9$ and $h = 0.635$ mm

First, set the initial specifications: $Z_0 = 50\ \Omega$, $f_c = 16$ GHz, $N = 2$, $|C(\omega)|_s = 0.707$, $B_w = 14 - 18$ GHz, $r = \pm 0.5$ dB. Then, choose the phase velocity, an initial coupler length, and step sizes for ω and l: $v = 104 \times 10^9$ mm/s, $l = 10\, l_c$, $\Delta l = 0.05$ mm, and $\Delta\omega = 0.1$. Specify a dummy channel: $|C_d(\omega)| = 0.3$ between 0 and 1 GHz. Compute $p_d(x)$ and $C(\omega)$ using eqs. (6.6) and (6.7), respectively. The design procedure given in Chapter 4 can be used for the rest of the computations.

The computed results are shown in Figure 6.3. The reflection coefficient distribution function, $p_d(x)$, for this design is shown in Figure 6.3(a). At two regions its value becomes negative for negative x values. The dummy channel compensates for the excess negative values thereby increasing $k(x)$ which is shown in Figure 6.3(b). The first peak of $k(x)$ corresponds to $p_d(x) = 0$ at the same point. Because $p_d(x)$ stays negative up to $x = -5.1$ mm, the corresponding $k(x)$ decreases. The amplitude of the dummy coupling determines the amplitude of $p_d(x)$ thence affecting the function $k(x)$. Therefore, the dips and peaks in $k(x)$ can be varied by changing the amplitude of the dummy channel. We can see from the computed physical dimensions (see $s(x)$) given in Figure 6.3(c) that a very low value of $k(x)$ at the dips may cause a large taper rate. Depending on the minimum desirable gap that corresponds to $k(0)$, $|C_d(\omega)|$ may be further increased or decreased to prevent large taper rates. The layout for this design is shown in Figure 6.3(d). This coupler is built on a $25.4 \times 25.4 \times 0.635$ mm^3 alumina substrate and its performance is measured using an HP8510C automatic network analyzer. The realized coupler is shown in Figure 6.4. The measured performance, which is not shown, was in good agreement with the computed results. The length of this coupler is 16.2 mm.

A second coupler with the same specifications but with $l = 3l_c$ and using the modification of $p(x)$ involving the multiplication of the even-mode impedance by a constant value is considered. The computed results for this coupler are shown in Figure 6.5. In this case $k(0)$ is 0.45, which is much higher than the first design. The coupler is 5.04 mm long.

The measured results for the second coupler are given in Figure 6.6. Very good agreement is observed between the measured and computed results. Input reflection is below -20 dB, and it includes reflections at the transitions. Measured isolation is around -30 dB, which confirms the effectiveness of wiggling and modification

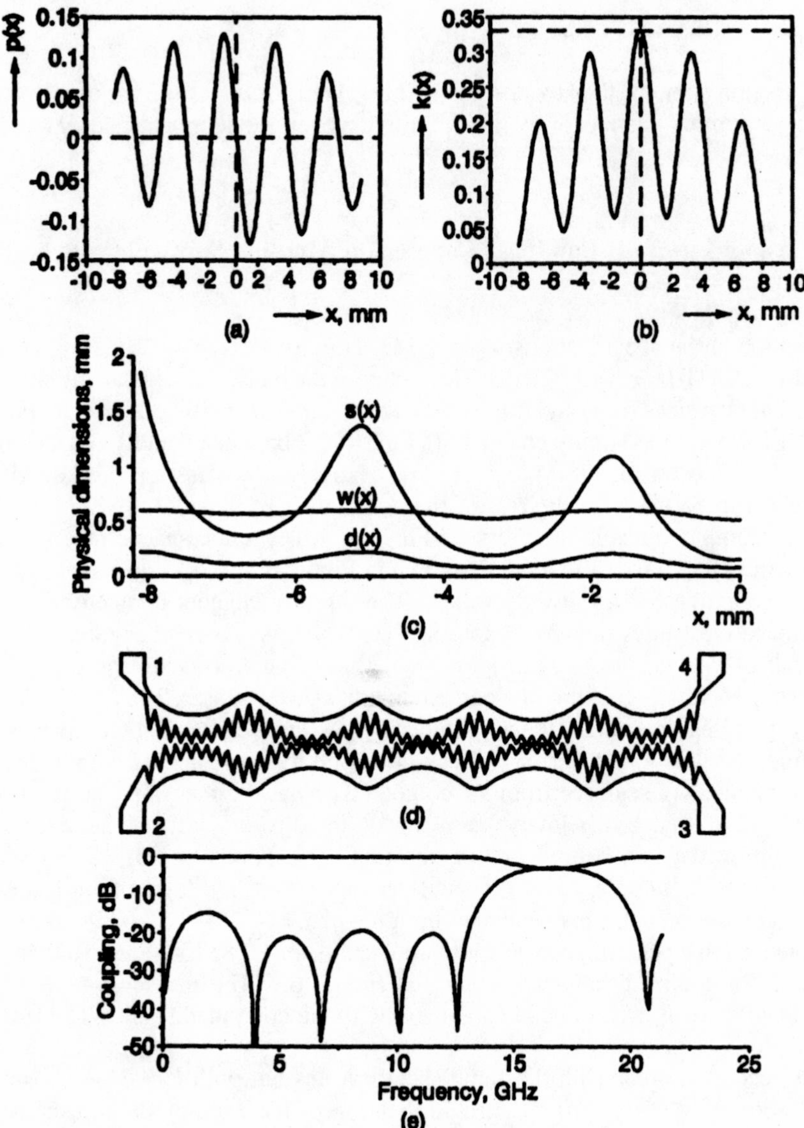

Figure 6.3 Computed results for a K_u-band, -3 dB bandpass coupler on alumina substrate with $\epsilon_r = 9.9$ and $h = 0.635$ mm: (a) modified reflection coefficient distribution function, (b) continuous-coupling coefficient, (c) continuous physical parameters, (d) bandpass coupler layout, and (e) performance.

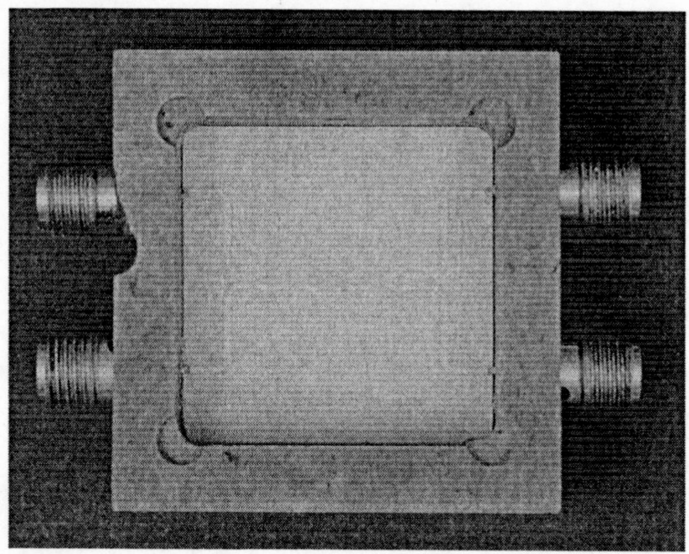

Figure 6.4 The K_u-band, 16.1-mm-long, -3 dB bandpass coupler.

technique for realizability. In the first design, isolation was measured to be 7 dB higher at the band edges, which is attributed to the sharper taper rate of the coupler. The deviation from phase quadrature in both cases was within $\pm 5°$ in the design bandwidth. For the smaller coupler this is shown in Figure 6.6(d).

These two designs show that, for a given bandwidth, almost identical performance can be obtained by using different modifications for $p(x)$. In practice, the choice of method largely depends on the system requirements. Further examples are given in the following sections.

6.4.2 *X*-band, -3 dB Bandpass Coupler on Alumina Substrate with $\epsilon_r = 9.9$ and $h = 0.635$ mm

First, set the initial specifications: $Z_0 = 50\ \Omega$, $f_c = 10$ GHz, $N = 2$, $|C(\omega)|_s = 0.707$, $B_w = 8$–12 GHz, $r = \pm 0.5$ dB. Then, choose the phase velocity, an initial coupler length, and step sizes for ω and l: $v = 108 \times 10^9$ mm/s, $l = 5l_c$, $\Delta l = 0.05$ mm, and $\Delta\omega = 0.1$.

After computing $p(x)$, the normalized even-mode impedance is multiplied by a constant ($E = 0.3$). The optimized continuous coupling coefficient is shown in Figure 6.7(a). Its maximum value is 0.44, which requires $w_c = 0.48$ mm and $s_c = 0.07$ mm at the center. The computed performance is shown in Figures 6.7(b) and (c).

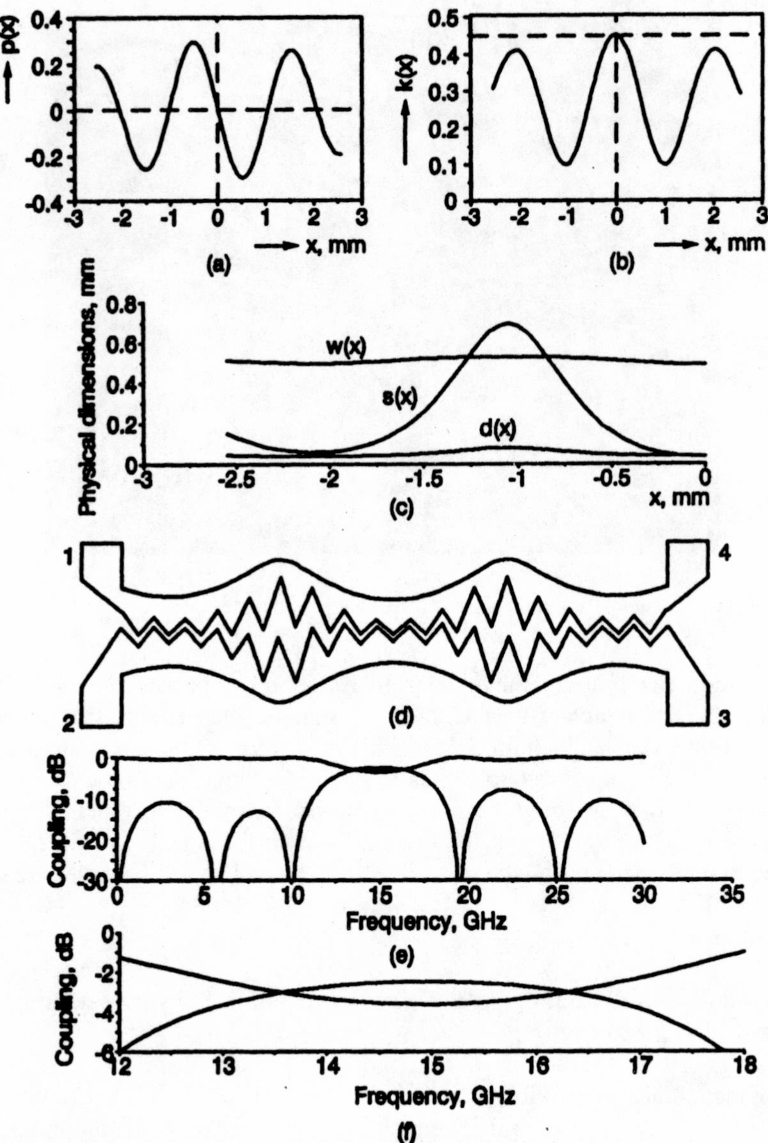

Figure 6.5 Computed results for a K_u-band, -3 dB bandpass coupler on alumina substrate with $\epsilon_r = 9.9$ and $h = 0.635$ mm: (a) modified reflection coefficient distribution function, (b) continuous-coupling coefficient modified by multiplying $Z_{0e}(x)$ by a constant, (c) continuous physical dimensions, (d) bandpass coupler layout, (e) performance in the specified frequency range, and (f) performance in the design band width.

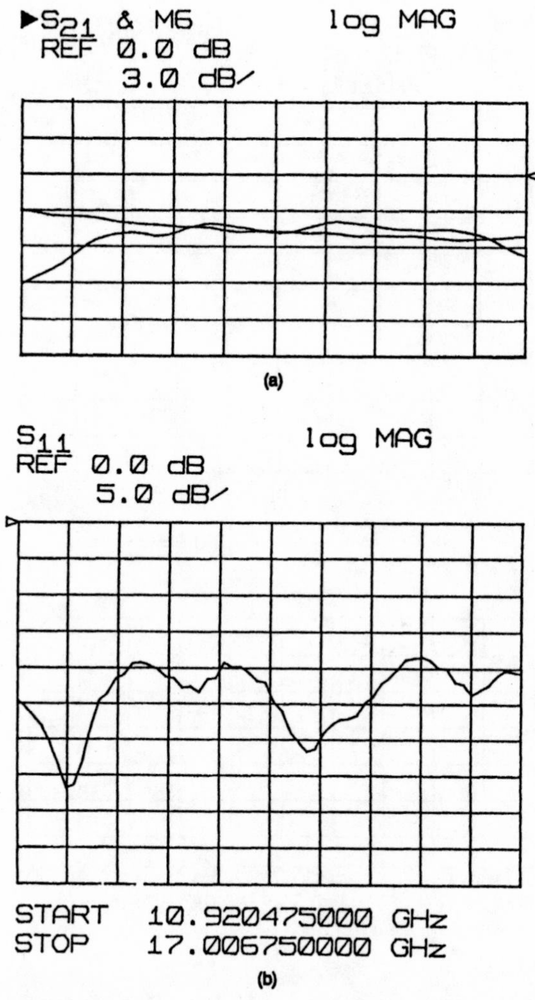

Figure 6.6 Measured results for the 5.04-mm-long, K_u-band, −3 dB coupler: (a) measured coupled and direct ports, (b) input reflection.

This coupler is built on a 25.4 × 25.4 × 0.635 mm^3 alumina substrate. Its measured performance is given in Figure 6.8. Good agreement is obtained with the computed results. The specified coupling is satisfied between 8.3 and 10.3 GHz bandwidth. Input reflection is kept below −20 dB. Isolation is around −35 dB in the design bandwidth. It is interesting to note the phase reversal in the measured phase quadrature, shown in Figure 6.8(d). This is caused by the network analyzer and has no physical meaning. Deviation from phase quadrature is ±2° in the design bandwidth.

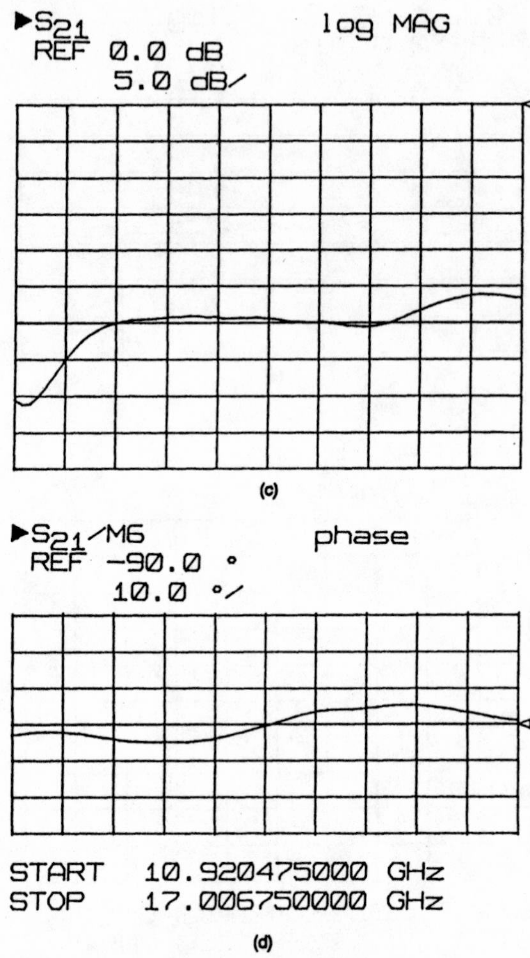

Figure 6.6 (continued) Measured results for the 5.04-mm-long, K_u-band, -3 dB coupler: (c) isolation, (d) deviation from phase quadrature.

In order to extend the usable bandwidth, it is necessary to increase the coupler length. The ripple amplitude significantly increases with the increasing constant E. This has been used advantageously to increase the coupling with single-lobe bandpass couplers. When the coupler length is increased, the bandpass coupling becomes similar to that of a multisection coupler (i.e., positive and negative ripples in the passband).

Consider a coupler with $l = 15\ \lambda c/4$ at 10 GHz. In this case realizability is achieved by specifying a dummy channel. The computed results are shown in Figure

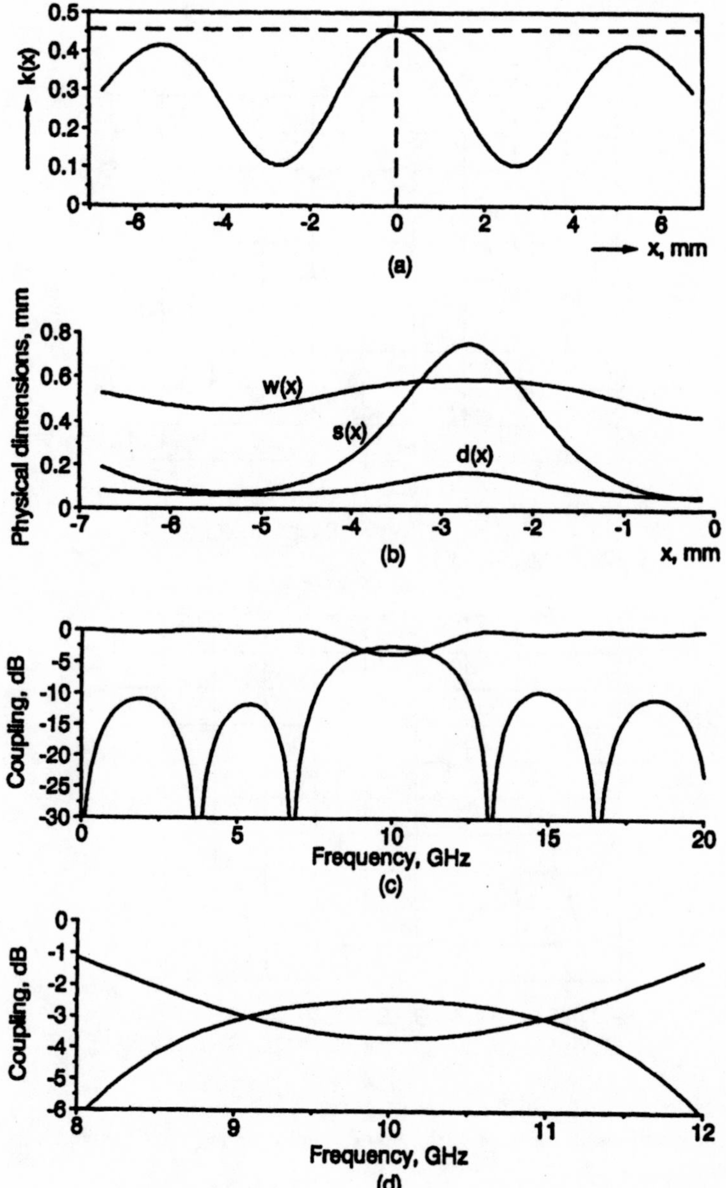

Figure 6.7 Computed results for an X-band, -3 dB bandpass coupler on alumina substrate with $\epsilon_r = 9.9$ and $h = 0.635$ mm: (a) continuous-coupling coefficient—modified by multiplying $Z_{0e}(x)$ by a constant, (b) continuous physical dimensions, (c) performance in the specified frequency range, and (d) performance in the design bandwidth.

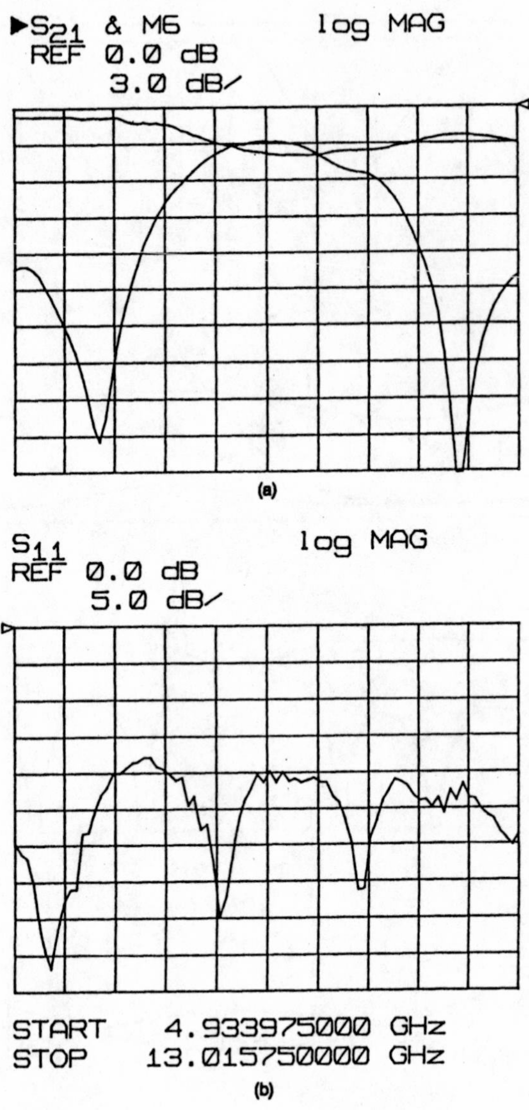

Figure 6.8 Measured results for the *X*-band, − 3 dB coupler: (a) measured coupled and direct ports, (b) input reflection.

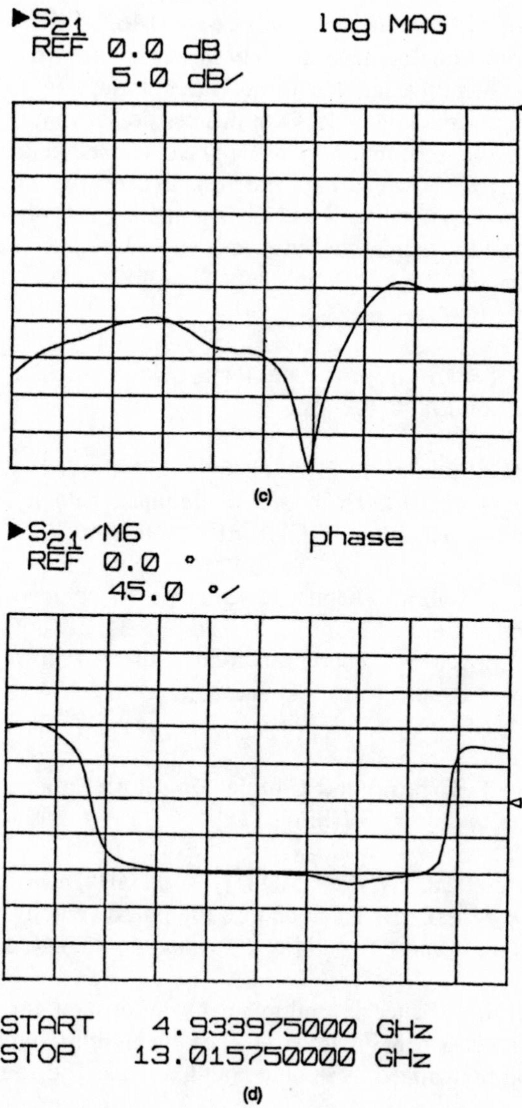

Figure 6.8 (continued) Measured results for the *X*-band, −3 dB coupler: (c) isolation, (d) deviation from phase quadrature.

6.9. The maximum coupling coefficient in this case is 0.462, which requires $w_c = 0.42$ mm and $s_c = 0.052$ mm with double-coupled lines. The computed performance shown in Figure 6.9(b), has increased number of nulls (zero couplings) in the out-of-band regions and has sharper band edges. It is not difficult to see that it looks more like a filter with 3 dB passband insertion loss. This will be discussed in the next chapter. The design band of interest is shown in Figure 6.9(c). Because we have introduced two discontinuities at $f = 8$ GHz and $f = 12$ GHz, coupling roll-off at these regions is high. Therefore, a slightly increased bandwidth should be specified to achieve the desired operational band.

6.4.3 K_a-band, −8.34 dB Bandpass Coupler on GaAs Substrate with $\epsilon_r = 12.9$ and $h = 0.2$ mm

First, set the initial specifications: $Z_0 = 50\ \Omega$, $f_c = 33$ GHz, $N = 2$, $|C(\omega)|_s = 0.384$, $B_w = 28$–38 GHz, $r = \pm 1$ dB. Then, choose the phase velocity, an initial coupler length, and step sizes for ω and l: $v = 94 \times 10^9$ mm/s, $l = 5l_c$, $\Delta l = 0.03$ mm, and $\Delta\omega = 0.1$.

In this design, a dummy channel is specified to achieve realizability. The computed results are given in Figure 6.10. The −8.34 dB coupling can also be achieved by using uniform wiggly couplers, which require a uniform gap of 0.038 mm. In the case of bandpass couplers this gap is increased to 0.094 mm.

6.4.4 75–95 GHz, −3 dB Bandpass Coupler on GaAs Substrate with $\epsilon_r = 12.9$ and $h = 0.1$ mm

First, set the initial specifications: $Z_0 = 50\ \Omega$, $f_c = 85$ GHz, $n = 2$, $|C(\omega)|_s = 0.707$, $B_w = 75$–95 GHz, $r = \pm 1$ dB. Then, choose the phase velocity, an initial coupler length, and step sizes for ω and l: $v = 94 \times 10^9$ mm/s, $l = 5\lambda c/4$, $\Delta l = 0.01$ mm, and $\Delta\omega = 0.1$.

The even-mode impedance is multiplied by a constant for realizability. The computed results are shown in Figure 6.11. The maximum coupling coefficient is 0.417 mm, which can be realized by double-coupled lines. The coupler length in this case is 1.36 mm.

With the preceding examples, we have tried to highlight the importance and wide applicability of bandpass couplers at both microwave and millimeter wave frequencies. The choice of method for realizability depends largely on the technology, bandwidth, and minimum and maximum desirable gaps. These couplers have been demonstrated to have high directivities, medium lengths, and moderate bandwidth.

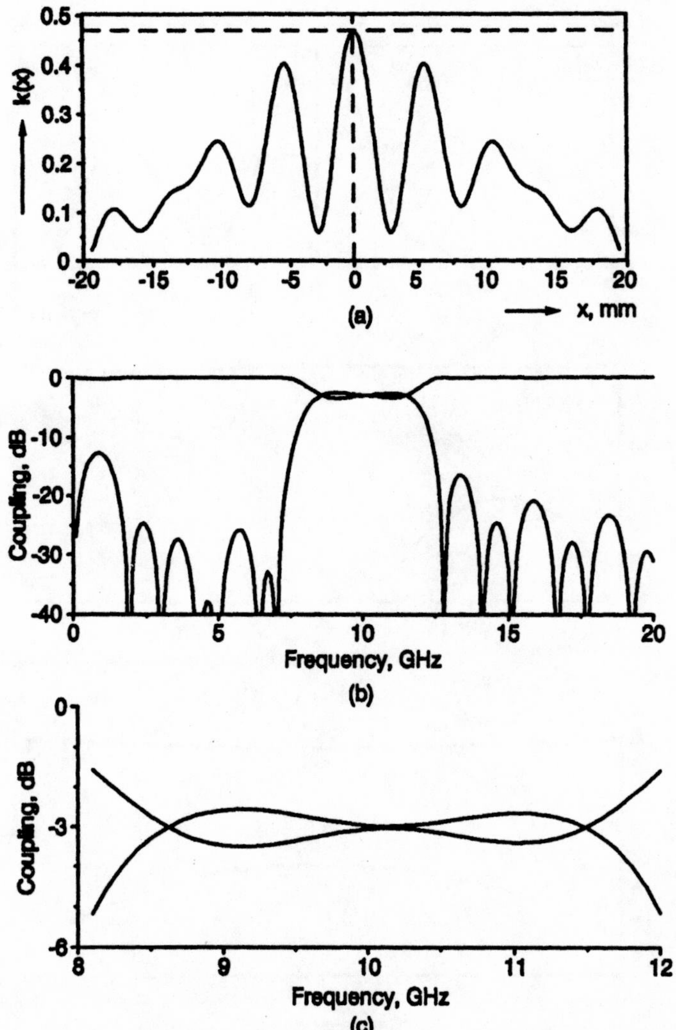

Figure 6.9 Computed results for the 40-mm-long, *X*-band, −3 dB coupler on alumina substrate with $\epsilon_r = 9.9$ and $h = 0.635$ mm: (a) continuous-coupling coefficient—modified by introducing a dummy coupling channel, (b) performance in the specified frequency range, and (c) performance in the design bandwidth.

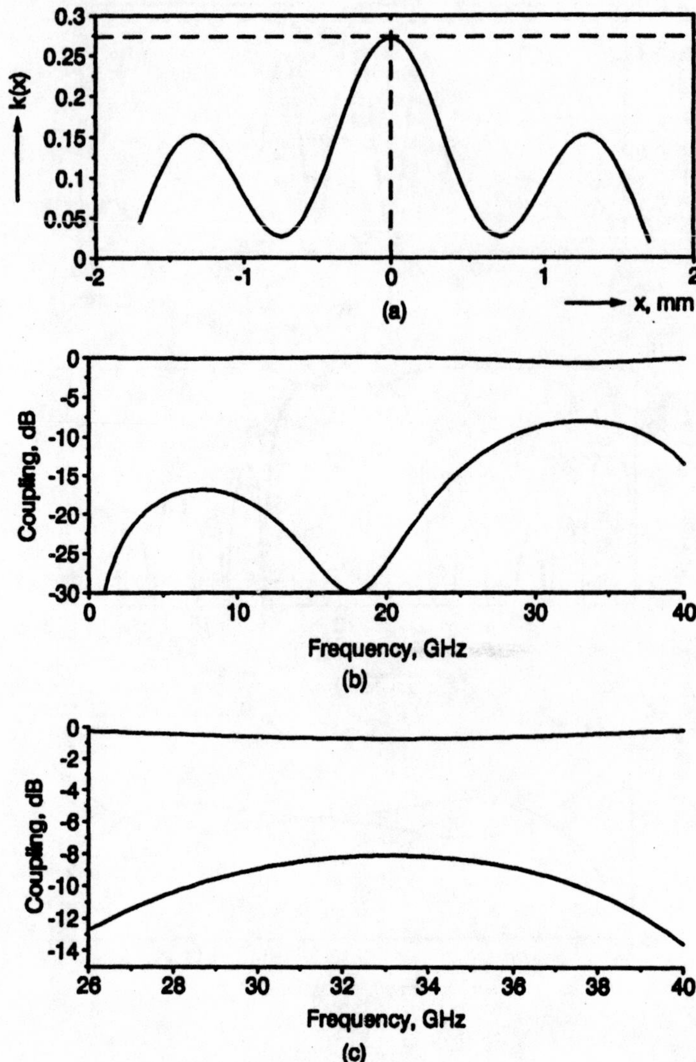

Figure 6.10 Computed results for a K_a-band, -8.34 dB bandpass coupler on GaAs substrate with $\epsilon_r = 12.9$ and $h = 0.2$ mm: (a) continuous coupling coefficient—modified by introducing a dummy coupling channel, (b) performance in the specified frequency range, and (c) performance in the design bandwidth.

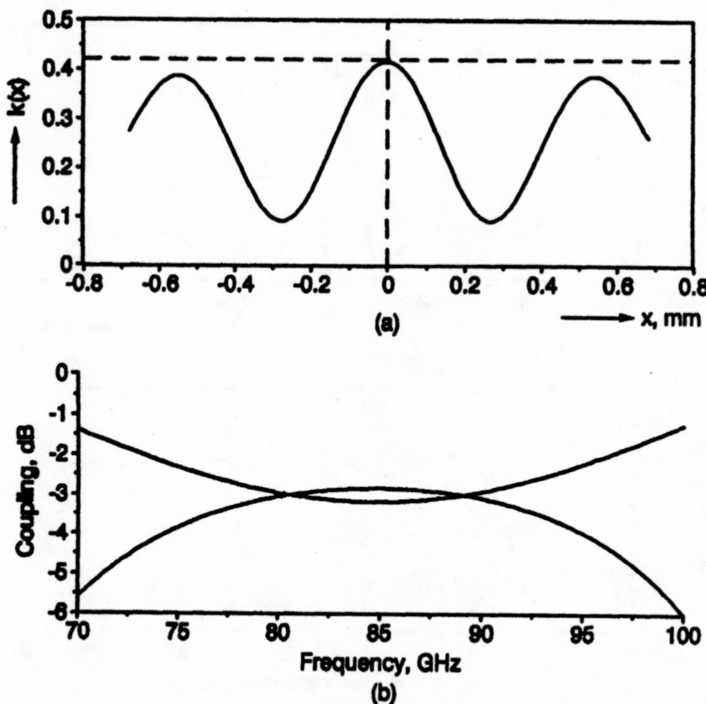

Figure 6.11 Computed performance for a 75–95 GHz, −3 dB bandpass coupler on GaAs substrate with $\epsilon_r = 12.9$ and $h = 0.1$ mm: (a) continuous-coupling coefficient—modified by multiplying even-mode impedance by a constant and (b) performance in the design bandwidth.

6.5 MULTIOCTAVE BANDPASS COUPLERS

We have seen in Chapter 4 that to realize multioctave band performance we need couplers with an interdigitated center section for nominal couplings between −14 and −7 dB. In the case of bandpass couplers, the need for an interdigitated section is eliminated by modifying $p(x)$. This will be illustrated by a computed example for a −10 dB coupler in the 10–40 GHz frequency range. For this example, we choose alumina substrate with $\epsilon_r = 9.9$ and $h = 0.25$ mm and an impedance level of 50 Ω. Realizability is achieved by specifying a dummy channel. The computed results are shown in Figure 6.12. The coupling coefficient has a maximum value of 0.44, which requires $w_c = 0.16$ mm and $s_c = 0.022$ mm with double-coupled lines. Because the required nominal coupling is not high, the dummy coupling amplitude is kept at a low value (−20 dB). The coupler length is 13.1 mm, which is reasonable for a 4:1 bandwidth.

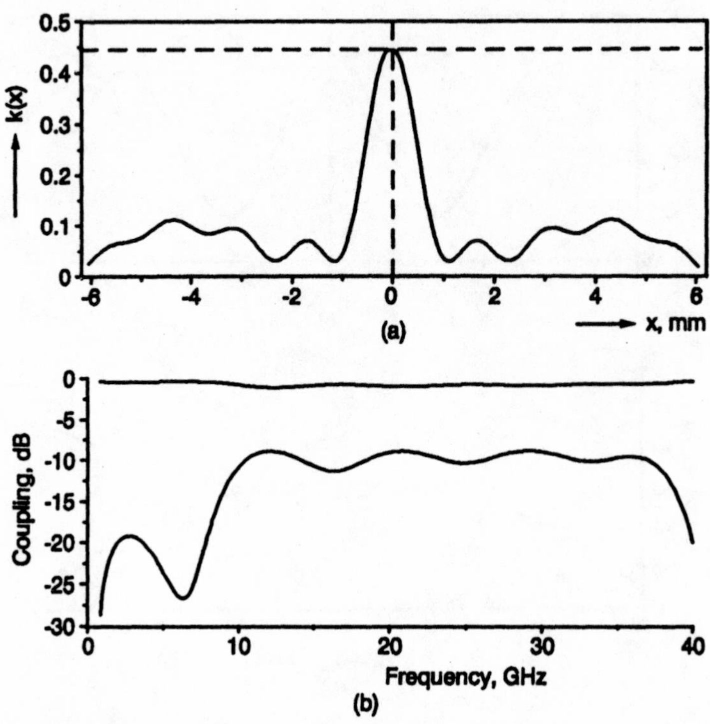

Figure 6.12 Computed performance for a 10–40 GHz, −10 dB coupler on alumina substrate with $\epsilon_r = 9.9$ and $h = 0.25$ mm: (a) continuous-coupling coefficient—modified by introducing a dummy coupling channel and (b) performance in the specified frequency range.

For a tight (e.g, −3 dB) multioctave coupling with bandpass couplers, we require an interdigitated section. This, still, is a significant improvement over the conventional design because the need for a tandem connection is eliminated. This is illustrated with an 8–20 GHz, −3 dB coupler again on alumina substrate with $\epsilon_r = 9.9$ and $h = 0.635$ mm. The computed results for this design are shown in Figure 6.13. An interdigitated section with length 1.6 mm is required at the bandpass coupler center. The maximum coupling coefficient in this case is 0.72, which requires $w_c = 0.069$ mm and $s_c = 0.047$ mm with $N = 4$. The total coupler length is 24.2 mm.

6.6 MULTI-LEVEL COUPLERS

Whether intentional or unintentional, the coupling response has several coupling lobes (the number of lobes depends on the coupler length) with different amplitudes.

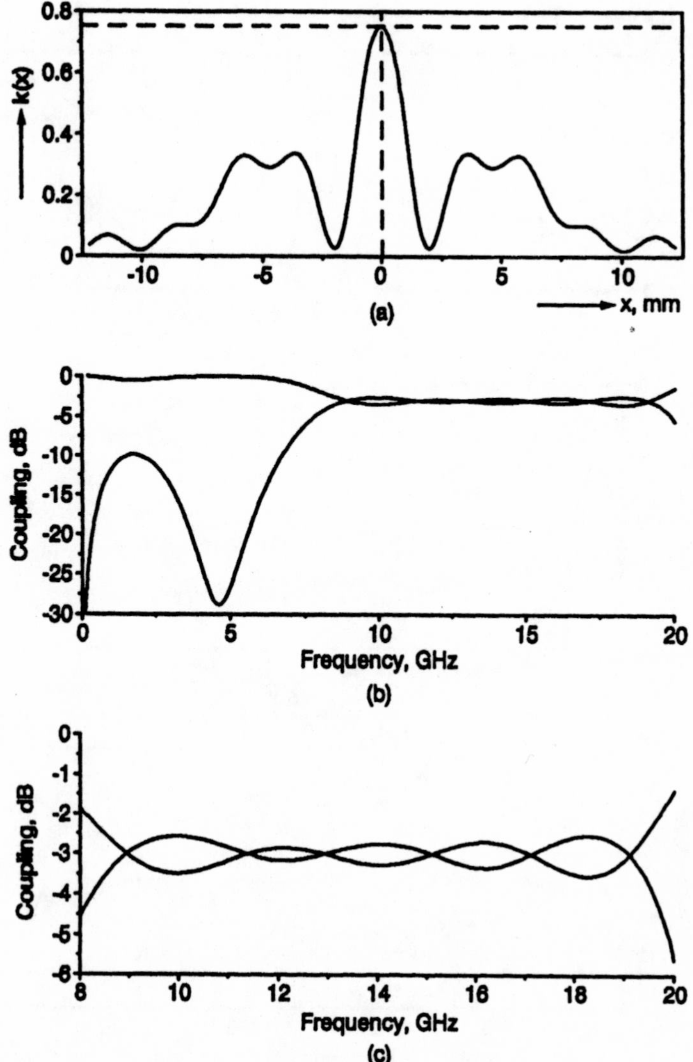

Figure 6.13 Computed performance for an 8–20 GHz, −3 dB coupler on alumina substrate with $\epsilon_r = 9.9$ and $h = 0.635$ mm: (a) continuous-coupling coefficient—modified by introducing a dummy coupling channel, (b) performance in the specified frequency range, and (c) performance in the design bandwidth.

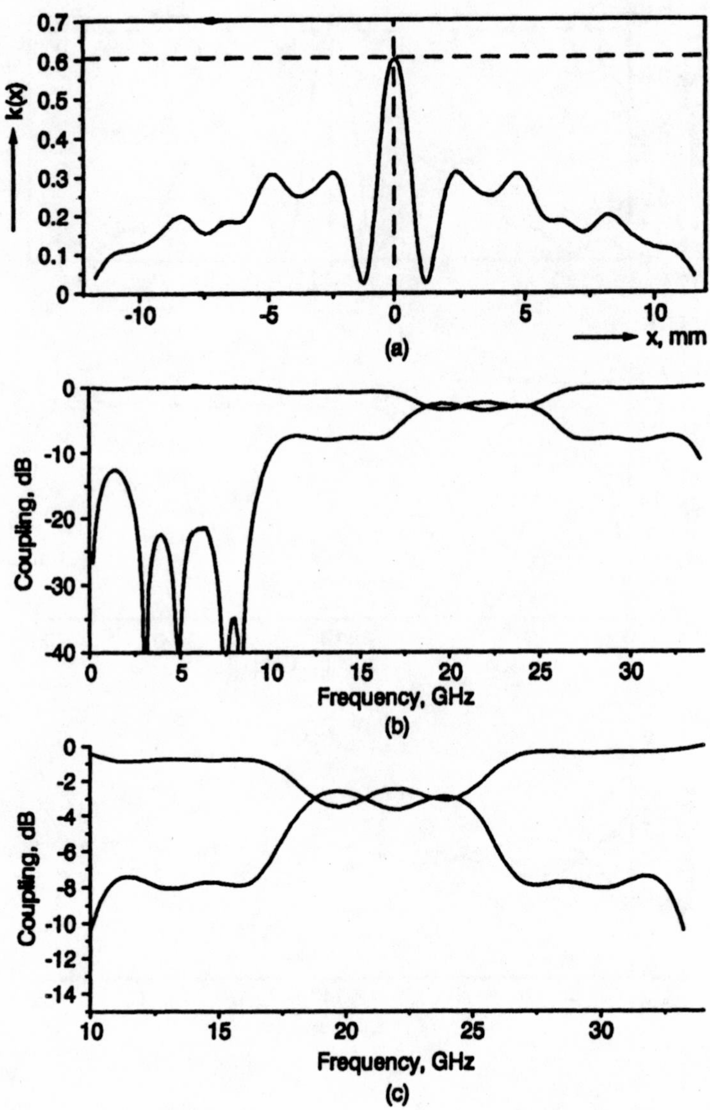

Figure 6.14 Computed performance for a 10–34 GHz, −8.34 to −3 to −8.34 dB multilevel coupler on alumina substrate with $\epsilon_r = 9.9$ and $h = 0.5$ mm: (a) continuous-coupling coefficient—modified by introducing a dummy coupling channel, (b) performance in the specified frequency range, and (c) performance in the design bandwidth.

Nonzero coupling values can be specified to deliberately change the coupling levels. The nominal coupling values in specified bandwidths can be maintained at desired values by taking a sufficiently long coupler. In principle, the design of multilevel couplers is similar to that of bandpass couplers given in the previous sections.

A 10–34 GHz, −8.34 to −3 to −8.34 dB multilevel coupler is designed on alumina substrate with $\epsilon_r = 9.9$ and $h = 0.5$ mm to illustrate the principle of multilevel coupling. The computed results are given in Figure 6.14. We observe that a 1 mm long interdigitated section is required at the center. A high $k(0)$ was expected because the bandwidth is wide and the nominal couplings are tight. For narrower bandwidths or moderate couplings, the coupler can be realized using double-coupled lines only. Here, the main objective is to show various possible design approaches for bandpass couplers.

REFERENCES

[1] Kammler, D. W., "The Design of Discrete *N*-Section and Continuously Tapered Symmetrical Microwave TEM Directional Couplers," *IEEE Trans. on Microwave Theory and Tech.*, Vol. MTT-17, 1969, pp. 577–590.

[2] Uysal, S. and J. Watkins, "Novel Microstrip Multifunction Directional Couplers and Filters for Microwave and Millimeter-Wave Applications," *IEEE Trans. on Microwave Theory and Tech.*, Vol. MTT-39, 1991, pp. 977–985.

Chapter 7
Bandpass Nonuniform Line Directional Filters

7.1 INTRODUCTION

Two basic mechanisms are at work in designing filters: (1) reflection at the input and (2) directonal filters with matched characteristics. In the first type, the undesired signal is reflected, which makes it impossible to maintain a good match across the attenuation band for the filter. Lossy absorbing elements may be used to maintain a good input match in the attenuation band. This is highly desirable especially when the rejected signal amplitude is high. Directional filters, on the other hand, give matched performance in the entire frequency range (both attenuation bands and passband) of the filter. This is possible because the rejected signal band is directed to another port, which is terminated. Every active microwave device generates noise (harmonics) that may significantly degrade the performance of other devices in the system. The designer can prevent this by using the necessary filtering arrangements. In such applications the required filter parameters are known. Every received signal is accompanied by noise (interference from other sources). The noise is removed by the system designer by placing a filter at the receiver front end. In this case, the required filter parameters are not known because the exact nature of noise is difficult to predict. A general purpose filter with high-rejection, low-bandpass insertion loss and sharp skirts is required.

In general, filters are required to be low loss (high Q), lightweight, and small. Low-loss filters are required especially for space applications, where the available power is limited. Lightweight and a small size enables them to be used in numbers in microwave communication systems to compensate for the faults of other circuits. Planar microstrip filters have the added advantage that they can easily be integrated with other microwave devices on the same substrate or the subassembly.

It is, therefore, clear that a universal specification sheet for filters makes little sense. What we look for in a filter design is primarily its frequency range of applicability, ease in design, repeatability, and cost.

The bandpass coupling principles of the previous chapter can be extended to design bandpass filters. These components require a careful control of the coupling in a

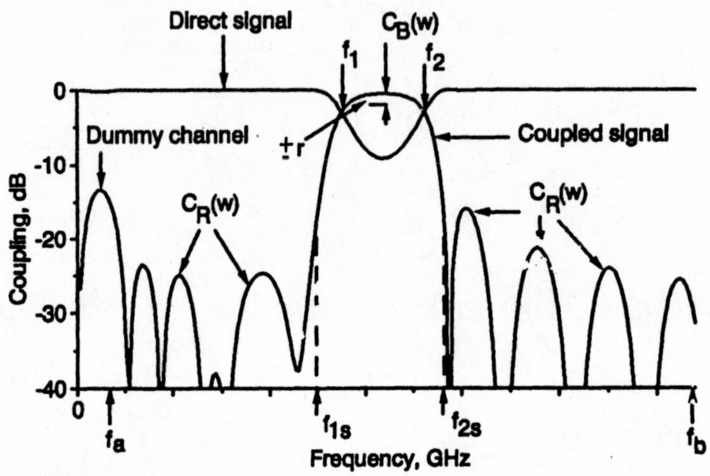

Figure 7.1 Bandpass filter parameters.

given frequency range. In this chapter, we investigate some of the key features of nonuniform coupled lines that apply to bandpass filters.

7.2 DESIGN PRINCIPLES

The modifications for the reflection coefficient distribution function suggested in Chapter 6 are equally applicable to bandpass filters. However, in the case of filters, we consider all the factors affecting the coupling in a given frequency range of interest. The choice of modification method, therefore, depends largely on the application. For example, we may need to reject a known frequency component with little attenuation on the desired signal. This can be done by adjusting the length of the filter so that a null coupling point corresponds to the unwanted frequency component. Alternatively, dummy channels can be specified to shift the finite out-of-band coupling positions without having to change the length of the filter. This will be discussed in a later section.

The bandpass filter principle using nonuniform bandpass couplers is shown in Figure 7.1. The filter parameters are:

- Bandpass coupling: $C_B(\omega)$
- Bandpass ripple: $\pm r$
- Bandwidth: from f_1 to f_2
- Frequency range: from f_a to f_b
- Rejection level: $C_R(\omega)$
- Bandpass skirt sharpness: $f_1 - f_{1s}$ and $f_{2s} - f_2$ for specified attenuation at f_{1s} and f_{2s}, respectively.

All these parameters can be related to the nonuniform coupled-line bandpass filter length and physical realizability of the coupling coefficient. Theoretically, or otherwise, it is impossible to realize full power coupling with edge-coupled backward-wave couplers. However, it is possible to realize a high coupling amplitude, on the order of 0.95 or higher. Such high coupling would obviously ensure minimum coupling loss in the passband. Bandpass skirt sharpness is controlled by the length of the filter.

7.3 MATCHED BANDPASS FILTERS

For matched filter performance, the coupled-arm response is taken as the filter output. The rest of the signal exits from the direct port, which may be terminated or used for other purposes. This filter is illustrated in Figure 7.2. The isolated port of the coupler-

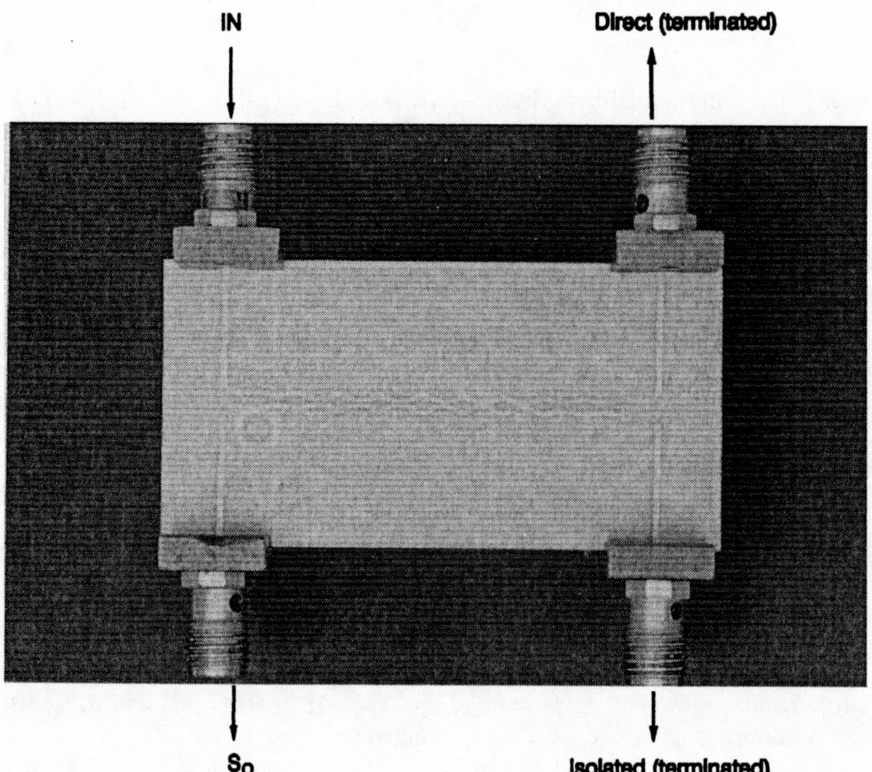

Figure 7.2 Bandpass filter.

filter is also terminated to eliminate reflections from this port. The output for this filter is given by

$$S_o = 20\log\left[\frac{|C(\omega)|}{1+\alpha(f)}\right] \tag{7.1}$$

where $\alpha(f)$ is the frequency-dependent total loss. $C(\omega)$ is the coupled signal given by

$$C(\omega) = \tanh\left\{\int_0^l \sin(2\omega x/v)p_m(x)\,dx\right\} \tag{7.2}$$

where $p_m(x)$ is the modified reflection coefficient distribution function, which can be obtained by the modifications given in the previous chapter.

Several filters may be cascaded to increase rejection level. Assuming identical filters, the output becomes

$$S_o = 20n\log\left(\frac{|C(\omega)|}{1+\alpha(f)}\right) \tag{7.3}$$

where n is the number of filters. For $|C_B(\omega)| = 0.9$, $|C_R(\omega)| = 0.1$, $\alpha(f) = 0$, and $n = 2$; and assuming identical filters we have $|S_{o_B}| = 0.81$ and $|S_{o_R}| = 0.01$. A 10-fold decrease in the rejected signal amplitude with only 10% reduction in the desired signal level can be achieved by cascading two identical coupler-filters.

7.4 DESIGN EXAMPLES

We have seen in Chapter 6 that a high out-of-band coupling means a higher coupling coefficient at the coupler center. Because we require a high coupling for minimum passband insertion loss, it is necessary to reduce the excess coupling so that the synthesized $k(x)$ is realizable with double-coupled lines. This requirement is met by increasing the coupler length.

7.4.1 *X*-band Bandpass Filter on Alumina Substrate with $\epsilon_r = 9.9$ and $h = 0.635$ mm

Set the initial specifications: $Z_0 = 50\ \Omega$, $f_c = 10$ GHz, $N = 2$, $|C_B(\omega)|_s = 0.95$, $B_w = 9$–11 GHz, $r = \pm 0.5$ dB. Choose phase velocity, initial coupler length, and step sizes for ω and l: $v = 108 \times 10^9$ mm/s, $l = 18l_c$, $\Delta l = 0.05$ mm, and $\Delta\omega = 0.1$. Specify a dummy channel: $|C_d(\omega)| = 0.25$ between 0 and 1 GHz.

The computed results are shown in Figure 7.3. The value of $k(0)$ is 0.426, which is easily realizable with double-coupled lines. The computed filter parameters are

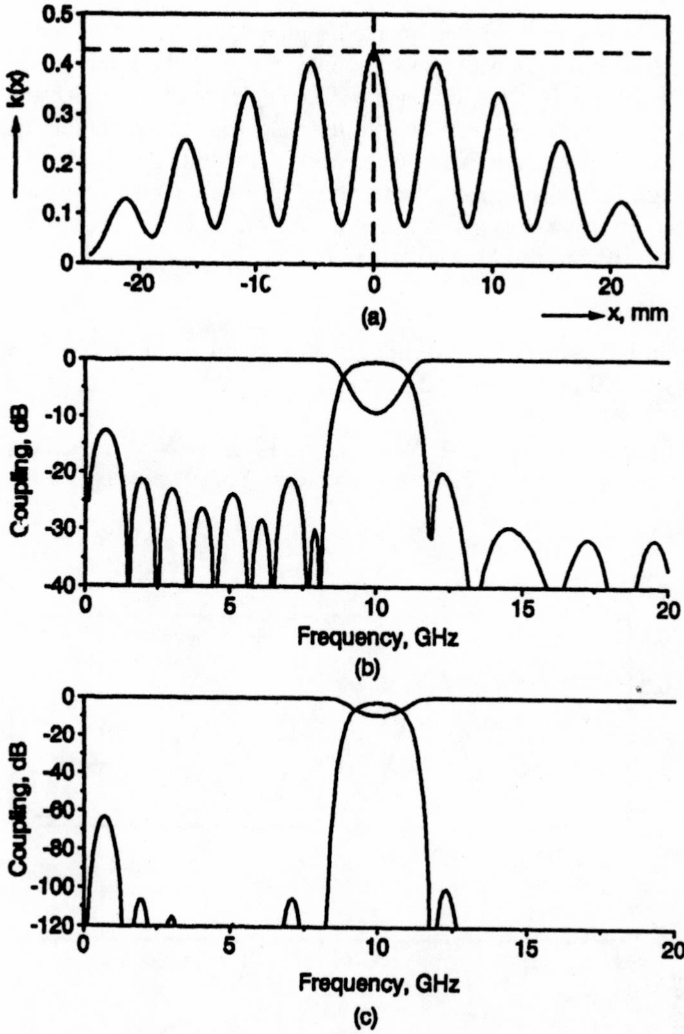

Figure 7.3 Computed results for a 9–11 GHz matched bandpass filter on alumina substrate: (a) continuous-coupling coefficient, (b) filter performance with $n = 1$, and (c) filter performance with $n = 5$.

$C_B(\omega) = -0.5$ dB, $r = \pm 0.2$ dB, $f_1(-3\text{ dB})$–$f_2(-3\text{ dB}) = 9$–11 GHz, f_a–$f_b = 2$–20 GHz. $C_R(\omega) = -20$ dB, $f_{1s}(-20\text{ dB}) = 9.5$ GHz, and $f_{2s}(-20\text{ dB}) = 11.5$ GHz.

We see from the coupling coefficient that there is enough room for a slight increase in $C_B(\omega)$ by specifying a higher coupling. It is also possible to vary the filter length for fine adjustment of the filter parameters. The dummy channel coupling amplitude that is higher than the rejected signal level is considered to be of no significance because it is outside the frequency range of the filter.

Further reduction in rejection level can be achieved at the expense of increased passband insertion loss (note that the ohmic, dielectric, and radiation losses are not taken into account because the main objective is to illustrate the inherent insertion loss due to coupling). For this purpose, we use eq. (7.3) with $n = 5$ and again $\alpha(f) = 0$. The computed performance is given in Figure 7.3(c). Rejection level in this case is in excess of -100 dB but the filter Q is reduced.

7.4.2 Wideband, Medium-Loss Bandpass Filters on Alumina Substrate with $\epsilon_r = 9.9$ and $h = 0.635$ mm

Set the initial specifications: $Z_0 = 50\ \Omega$, $f_c = 10$ GHz, $N = 2$, $|C_B(\omega)|_s = 0.707$, $B_w = 8$–12 GHz, and $r = \pm 0.5$ dB. Choose the phase velocity, an initial coupler

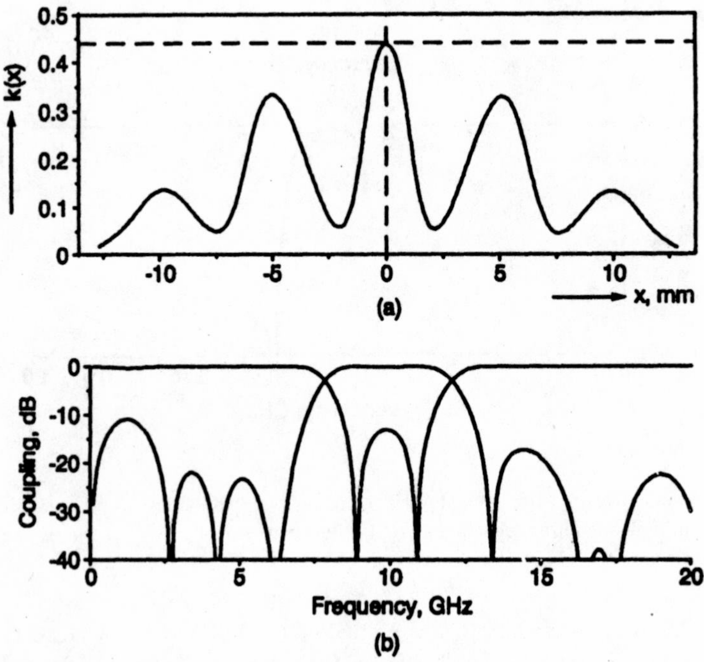

Figure 7.4 Computed results for an 8–12 GHz matched bandpass filter on alumina substrate: (a) continuous-coupling coefficient and (b) performance of tandem filter with identical coupler-filters.

length, and step sizes for ω and l: $v = 108 \times 10^9$ mm/s, $l = 12l_c$, $\Delta l = 0.05$ mm, and $\Delta\omega = 0.1$.

The computed results for this design are given in Figure 7.4. Because we have -3 dB coupling we can obtain a 0 dB coupler-filter by using tandem connection. However, the rejection level will also be increased with identical filters in tandem. The performance for the tandem connected filter is given in Figure 7.4 (b).

The passband insertion loss of the filter can be further increased to ease physical realizability. Variation in coupling coefficient decreases, thereby decreasing taper rates between coupling coefficient minima and maxima along the coupler length. The computed results for $C_B(\omega) = -10$ dB in X-band are given in Figure 7.5. We see from $k(x)$ that we have plenty of room for band widening with lossy filters. Figure 7.5(b) gives the computed performance for this filter.

A lossy bandpass filter with multioctave bandwidth is investigated and has $C_B(\omega) = -6$ dB. An 8–18 GHz band filter is realizable with double-coupled lines with the specified insertion loss. The computed performance for this design is given in Figure 7.5(c). It is indeed possible to employ interdigitated section or sections with higher bandpass coupling amplitudes; we shall see this in a later section.

7.4.3 K_u-Band Bandpass Filters on Alumina Substrate with $\epsilon_r = 9.9$ and $h = 0.25$ mm

Set the initial specifications: $Z_0 = 50\ \Omega$, $f_c = 16$ GHz, $N = 2$, $B_w = 14$–18 GHz, $r = \pm 0.2$ dB.

The first filter investigated has $C_B(\omega) = -0.5$ dB and $l = 30$ mm. The computed results for this filter are given in Figure 7.6. The $k(0)$ for this filter is 0.45, which may be a little high to realize with double-coupled lines on 0.25-mm-thick alumina. A second filter with 3 dB passband insertion loss is designed with the same initial specifications. The $k(0)$ shown in Figure 7.6(c) is 0.38, which is easier to realize with $N = 2$. We see that $k(x)$ is unnecessarily high toward the center, causing a higher out-of-band coupling in some regions. A second -3 dB coupler-filter may be connected in tandem for a theoretical 0 dB filter. Tandem filter performance is given in Figure 7.6(d).

The computed performance of a 10–14 GHz filter is given in Figure 7.7. A second dummy coupling channel is specified at a frequency much higher than the first passband. This dummy coupling channel, if desired, can be further shifted away from the main passband. Its amplitude can also be varied so that it does not influence a high $k(0)$. This type of filter has two major advantages: (1) the out-of-band signals between dummy channels and passband are further reduced, and (2) the number of coupling coefficient minima and maxima along the coupler are also reduced. The latter is particularly useful when the coupler-filter length is long. Details of such designs will be discussed in the next chapter.

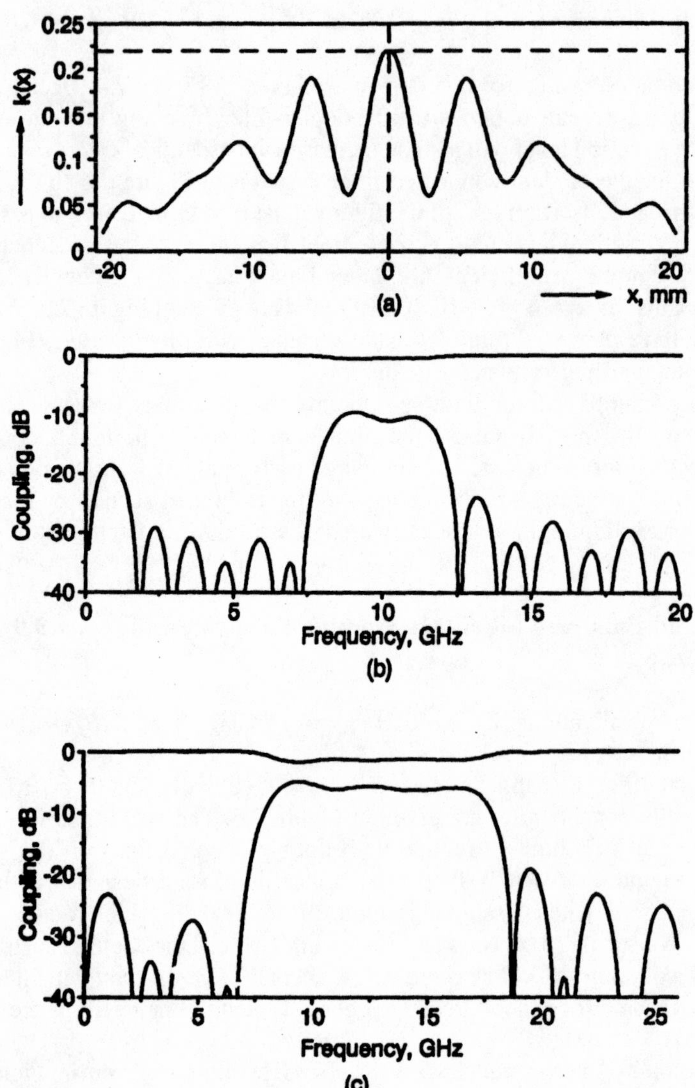

Figure 7.5 Computed results for lossy matched filters on alumina substrate: (a) continuous-coupling coefficient for a −10 dB coupler-filter, (b) filter performance in the 8–12 GHz bandwidth, and (c) −6 dB multioctave bandwidth filter performance.

7.4.4 Cascaded or Tandem Connected Filters with Nonidentical Designs

Consider a coupler-filter whose reflection coefficient distribution function is modified as follows:

$$\begin{aligned} p(x) = -\frac{2}{\pi v}\Bigg\{ &\int_0^{\omega_d} \sin(2\omega x/v)\tanh^{-1}[C_d(\omega)]\,d\omega \\ &+ \int_{\omega_d}^{\omega_1} \sin(2\omega x/v)\tanh^{-1}[C_1(\omega)]\,d\omega + \int_{\omega_1}^{\omega_2} \sin(2\omega x/v)\tanh^{-1}[C_2(\omega)]\,d\omega \\ &+ \cdots + \int_{\omega_{B_1}}^{\omega_{B_2}} \sin(2\omega x/v)\tanh^{-1}[C_B(\omega)]\,d\omega \\ &+ \int_{\omega_{B_2}}^{\omega_{B_m}} \sin(2\omega x/v)\tanh^{-1}[C_m(\omega)]\,d\omega \\ &+ \int_{\omega_{B_{m+1}}}^{\omega_{B_{m+2}}} \sin(2\omega x/v)\tanh^{-1}[C_{m+1}(\omega)]\,d\omega \\ &+ \cdots + \int_{\omega_{m+n}}^{2\omega_c} \sin(2\omega x/v)\tanh^{-1}[C_{m+n}(\omega)]\,d\omega \Bigg\} \end{aligned} \tag{7.4}$$

where $C_d(\omega)$ is the dummy channel for realizability and $C_B(\omega)$ is the passband coupling of the filter.

In the preceding equation, a number of dummy channels are specified on either side of the passband. The coupling functions other than $C_B(\omega)$ and $C_d(\omega)$ are the specified nonzero but small coupling amplitudes in the respective dummy channels. With two cascaded coupler-filters the dummy coupling bands are selected so that the maxima in the out-of-band regions correspond to the minima of the first coupler-filter and vice versa. This technique gives about 10 dB increased rejection compared to two identical cascaded or tandem-connected coupler-filter sections.

The method of specifying additional dummy coupling channels is very effective for selective optimization of out-of-band regions both for cascaded sections and for single coupler-filters. This is especially useful with reflection-type filters, which cannot easily be cascaded. A computed example with three additional dummy channels, specified above the passband and one below the passband of a K_u-band filter, is given in Figure 7.8. About 10 dB improvement is achieved for the upper stop band. This value is more than 10 dB just below the passband; however, the coupling amplitude in the first dummy channel was chosen to be a little higher than it should have been. The dummy channels can be chosen anywhere in the frequency range, and they may have negative or positive specified couplings.

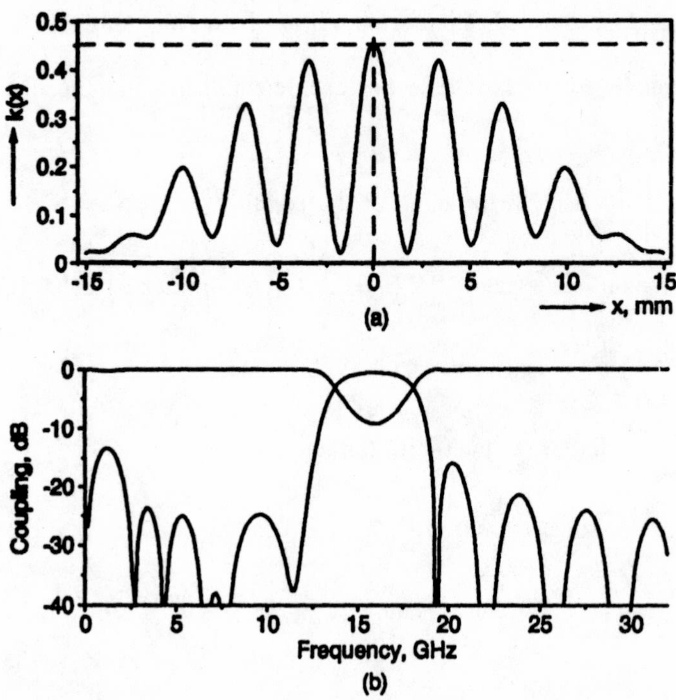

Figure 7.6 Computed results for K_u-band matched bandpass filters on alumina: (a) continuous-coupling coefficient for a -0.5 dB filter, (b) filter performance with $n = 1$.

7.4.5 Bandpass Filters with an Interdigitated Section

The passband insertion loss due to coupling of a single coupler-filter can be further reduced by increasing the specified coupling amplitude in the specified passband. We may also need to increase the bandwidth of the filter with minimum coupling loss. These requirements result in an increased $k(0)$ that is no longer realizable with double-coupled lines. With a moderate increase in filter bandwidth and specified passband coupling amplitude, it would be sufficient to use an interdigitated center section. With a further increase in filter bandwidth and specified passband coupling amplitude, the two coupling coefficient lobe amplitudes adjacent to the center will also be increased to such values, which are no longer realizable with double-coupled lines.

These two cases are illustrated with a 6–8 GHz coupler-filter design. Three examples are computed with $|C_B(\omega)| = 0.707$, 0.88, and 0.98 in the specified bandwidth with $l = 18l_c$. The computed continuous coupling coefficients are given in Figure 7.9. The design with $|C_B(\omega)| = 0.88$ requires an interdigitated center section for coupling coefficients between $k(-0.5)$ and $k(0.5)$. Design specifications for this

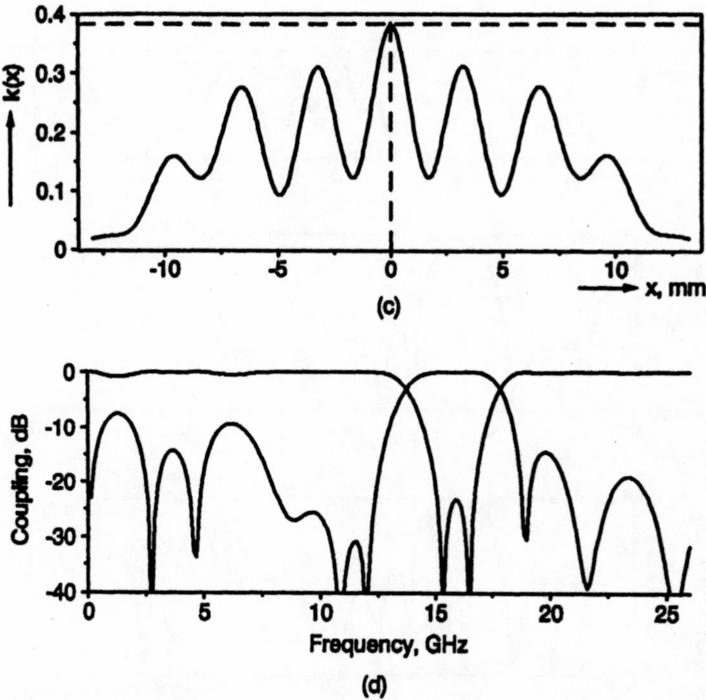

Figure 7.6 (continued) Computed results for K_u-band matched bandpass filters on alumina: (c) continuous-coupling coefficient for a -3 dB coupler-filter, (d) performance of the tandem filter with identical coupler-filters.

section can be obtained from Chapter 3. Depending on other requirements, anyone of the configurations given in Chapter 5 can be chosen for this section.

For the second case we keep the same bandwidth but specify a higher coupling amplitude in the passband. The coupling coefficient lobes adjacent to the center are increased to a maximum value of 0.7 and $k(0)$ to 0.75. We now need three interdigitated sections to realize the coupling coefficients at these regions. Because the lengths of these sections are short, they will have a negligible effect on the filter performance. However, the design complexity increases and the layout takes more time to generate.

The computed performances given in Figure 7.10, for these filters, show that, with a fixed-length coupler-filter, bandwidth increases with increasing passband coupling amplitude. The increase in out-of-band couplings is caused by an increase in the dummy coupling amplitudes for the two cases. By properly adjusting the dummy coupling amplitude, $k(0)$ values shown in Figure 7.9(b) and (c) can be reduced to 0.5 and 0.65, respectively. By reducing $k(0)$ to its lowest possible value, out-of-band couplings can be kept below -25 dB in both cases.

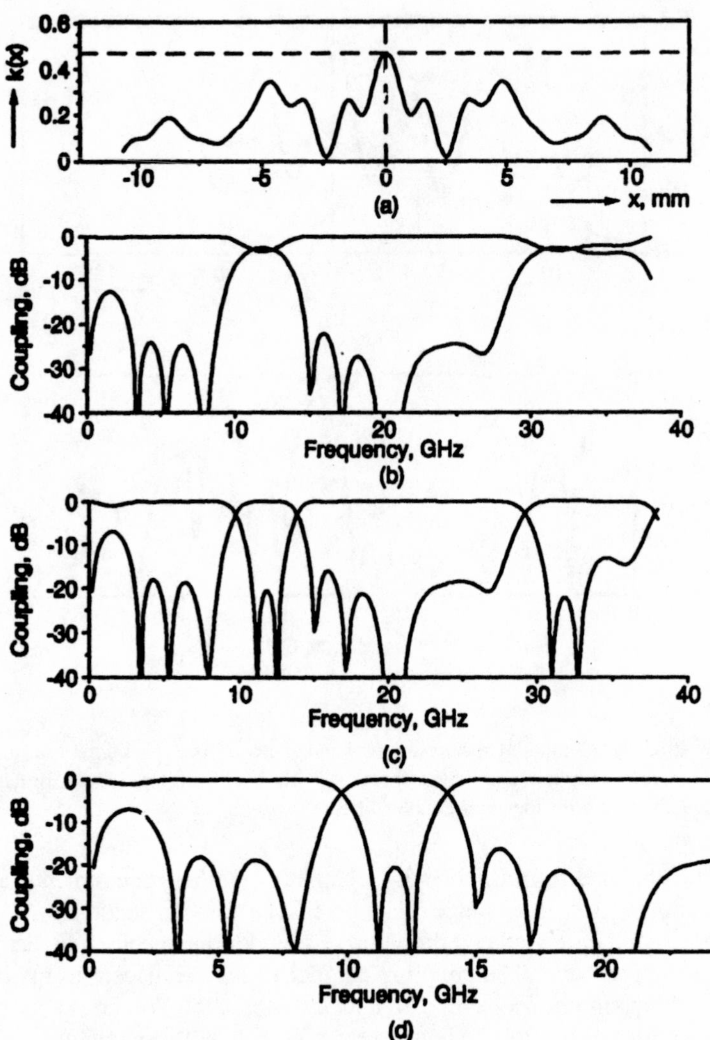

Figure 7.7 Computed results for a matched filter realized with an additional dummy coupling channel: (a) continuous-coupling coefficient, (b) −3 dB coupler filter performance, (c) tandem performance, and (d) filter performance.

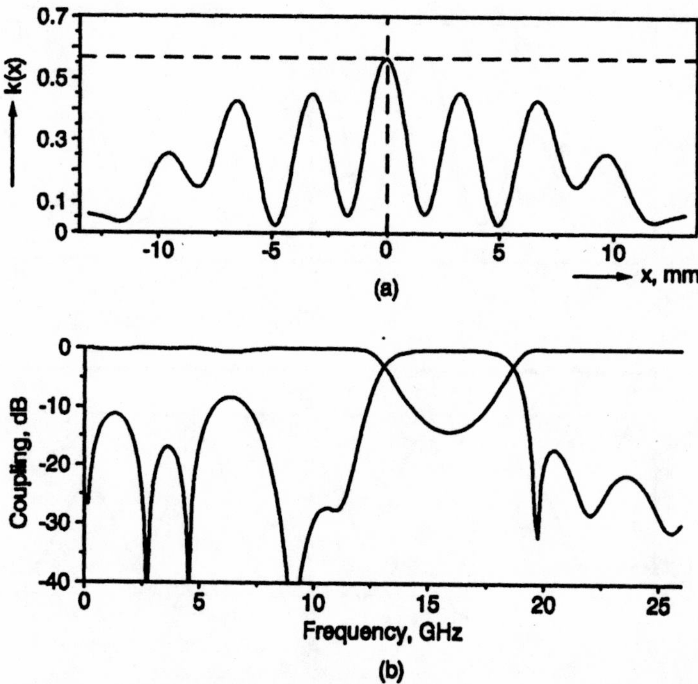

Figure 7.8 Computed results of a K_u-band matched filter: (a) continuous-coupling coefficient and (b) filter performance with additional dummy coupling channels: $C_1(\omega) = 0.2$ (5–7 GHz), $C_2(\omega) = -0.03$ (20–22 GHz), $C_3(\omega) = -0.03$ (22–24 GHz), and $C_4(\omega) = -0.03$ (24–26 GHz).

7.5 REFLECTION-TYPE BANDPASS FILTERS

Filters with reflection at the input can be designed by employing -3 dB bandpass coupler-filters. The design principle for these filters is illustrated in Figure 7.11. The coupled and direct ports are open circuited, and the signals reflected from these ports combine at the isolated port with S_o given by

$$|S_o| = 2|C(\omega)|\sqrt{1 - |C(\omega)|^2} \tag{7.5}$$

The reflected signal is obtained by

$$|S_{o_r}| = 1 - 2|C(\omega)|^2 \tag{7.6}$$

This technique ensures minimum passband insertion loss, and the bandpass ripple amplitude can be maintained at a desired value by adjusting the coupling balance.

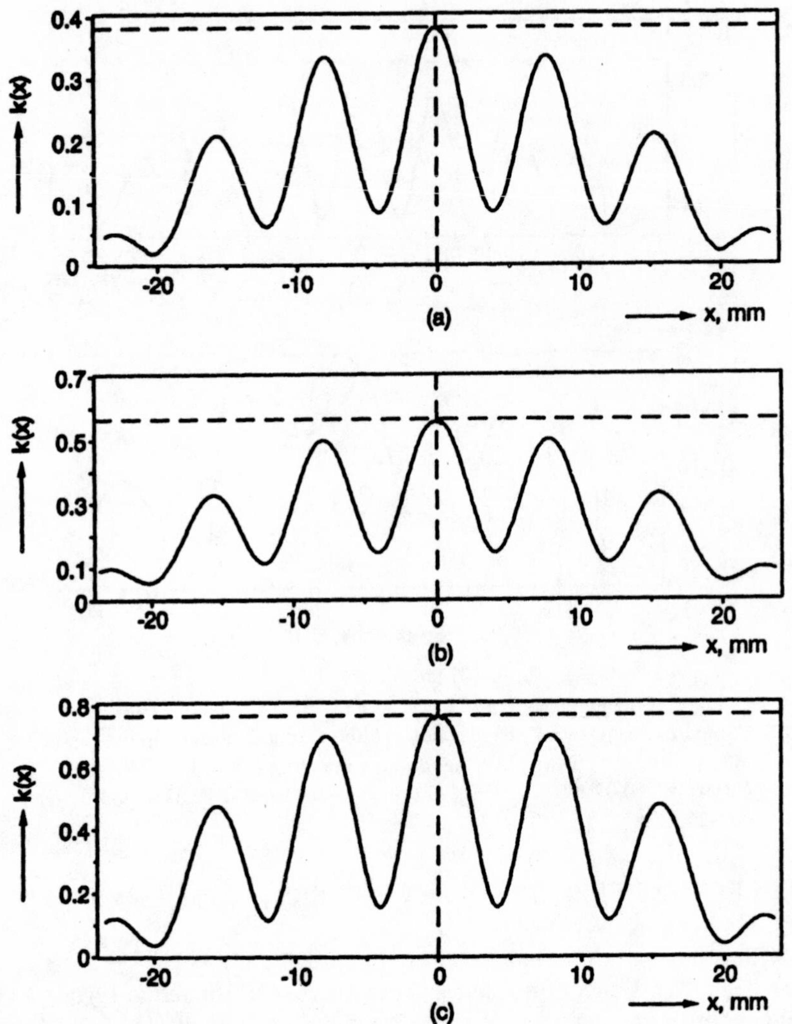

Figure 7.9 The computed continuous-coupling coefficients for X-band matched filters with: (a) $|C_B(\omega)| = 0.707$, (b) $|C_B(\omega)| = 0.88$, and (c) $|C_B(\omega)| = 0.98$.

The amplitude of the rejected signal is approximately twice of the out-of-band coupling. This rejection level can be improved by adding phase-shifting elements to the direct or coupled ports or both. With narrowband filters it is sufficient to add a half-wavelength-long line either at the coupled port or direct port. The reflected signal will be phase shifted by an amount $2\beta(\lambda_c/2)$ or $2\pi f/f_c$. With this technique it is sufficient to minimize the out-of-band coupling at the vicinity of the passband. The regions where out-of-band coupling is high can be selected to correspond to $f_c/2$ and $3\,f_c/2$. At these

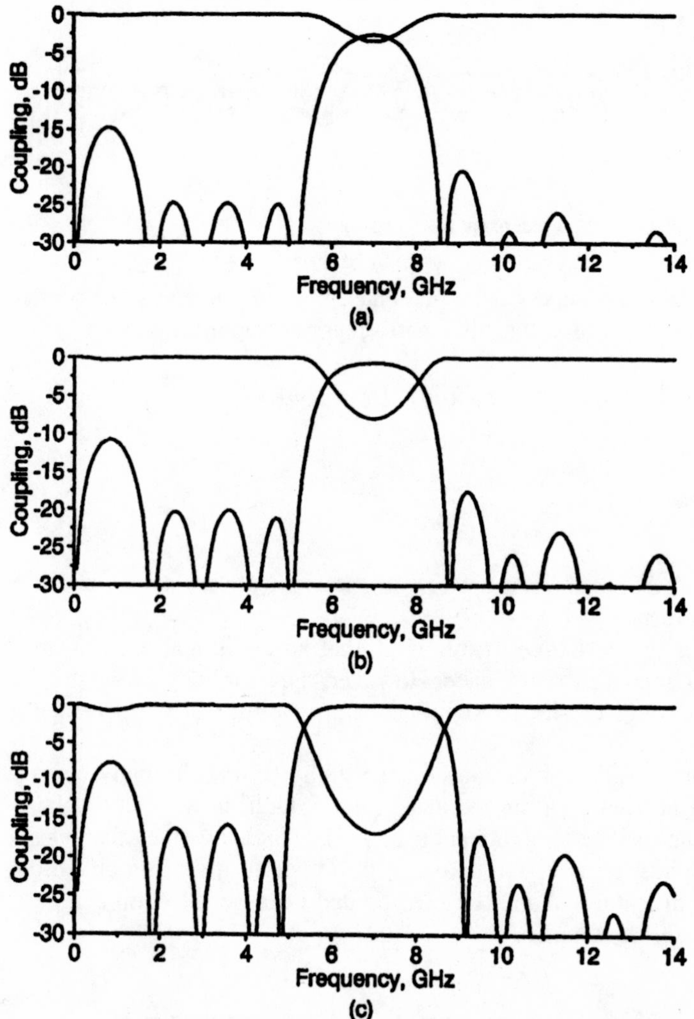

Figure 7.10 The computed results for *X*-band matched filters with: (a) $|C_B(\omega)| = 0.707$, (b) $|C_B(\omega)| = 0.88$, and (c) $|C_B(\omega)| = 0.98$.

points the two signals appearing at the isolated port of the coupler-filter will be out-of-phase with the addition of a half-wavelength-long line. This technique can also be used to design extremely narrowband filters by increasing the length of the added line.

With wideband filters we must further modify the phase-shifting element(s) so that the passband is not affected. The desired phase shift is shown in Figure 7.12. Short lengths of open-circuited lines and stubs would be sufficient to achieve the desired

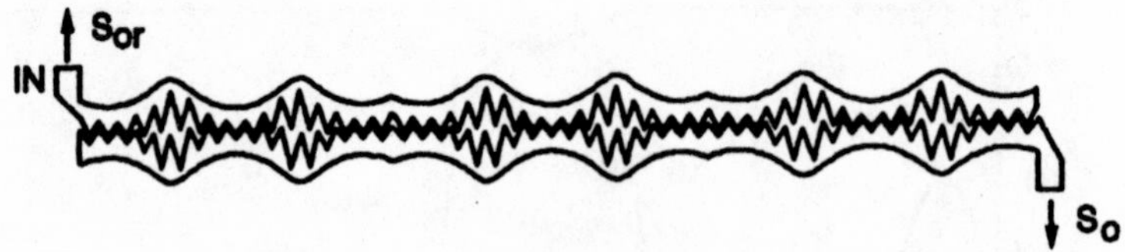

Figure 7.11 Reflection-type bandpass filter principle.

phase shift. In the general case when phase-shifting elements are connected to both direct and coupled ports, the filter performance is computed from

$$S_o = -jC(\omega)\sqrt{1 - C^2(\omega)}\,(e^{j\phi_1} + e^{j\phi_2}) \tag{7.7}$$

and

$$S_{o_R} = C^2(\omega)\,e^{j\phi_1} - j[1 - C^2(\omega)]\,e^{j\phi_2} \tag{7.8}$$

where ϕ_1 and ϕ_2 are the phase shifts of the elements connected to the coupled and direct ports, respectively.

It is for the first time in this book that an additional external modification is suggested to improve performance. However, the simplicity of the proposed external modification coupled with its significant improvement in performance justifies this effort.

So many modifications can be done with $p(x)$ that in many cases the required solution can be found within the design itself. Until now, with bandpass filters, we have used the simplest modification of $p(x)$, which involves the introduction of a dummy coupling channel for realizability. Other forms of modification, especially selective modification, should be investigated because rejection can be significantly increased at selected bands.

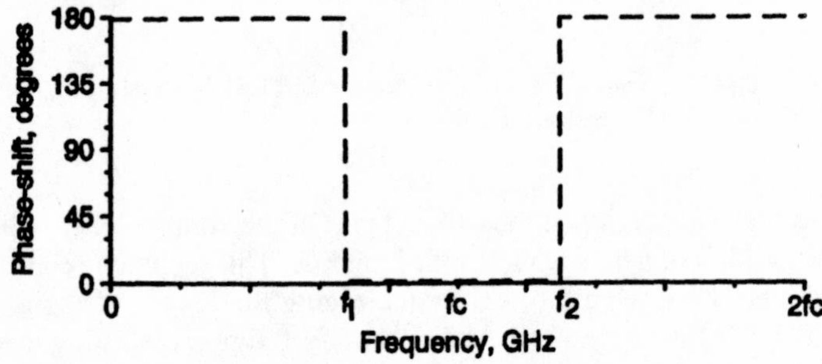

Figure 7.12 Ideal phase shift required for increased rejection level.

7.6 DESIGN EXAMPLES

In this section we shall give a few examples to illustrate bandpass filter design based on the reflection at the input. Because it is important to maintain a good input match at the design bandwidth, only −3 dB bandpass coupler-filters will be considered. It may, of course, be possible that a good input match is not required at all with some applications where coupler-filters other than −3 dB may be employed but this is left to the reader.

7.6.1 K_a-Band Bandpass Filter on GaAs Substrate with $\epsilon_r = 12.9$ and $h = 0.2$ mm

The first example in this section is given for a 41% bandwidth bandpass filter. We must maintain input reflection below −20 dB in the design bandwidth. This requires a coupling imbalance of ±1 dB from 26.5 to 40 GHz. This specification cannot be achieved with a single-lobe coupling in the design bandwidth. The coupler-filter length should be sufficiently long to achieve such a performance.

The desired performance is achieved by an 8.5-mm-long coupler-filter. The computed performance is given in Figure 7.13. The passband ripple is less than ±0.1 dB.

As expected the rejection level of this filter is low; it can be improved by selective minimization (specifying dummy channels) of the out-of-band coupling or simply by using a longer coupler-filter. Alternatively, simple passive phase-shifting elements can be connected to the coupled and direct ports.

The length of this filter may be long for MMIC realization. Folding the filter should be possible at the tightly coupled regions. A symmetrical folding, that is, an equal number of bends for direct and coupled lines, is recommended to maintain phase quadrature in the filter passband.

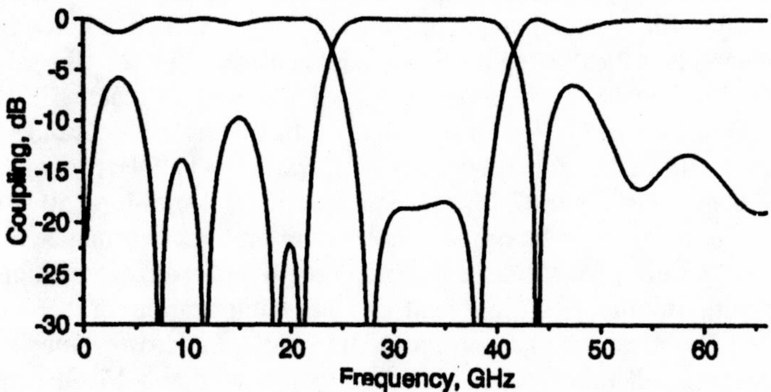

Figure 7.13 Computed performance for a K_a-band, reflection-type bandpass filter on GaAs.

7.6.2 K_u-Band Bandpass Filter on Alumina Substrate with $\epsilon_r = 9.9$ and $h = 0.25$ mm

The filter passband in this case is selected as 14–18 GHz. A single-lobe −3 dB bandpass coupler is sufficient to maintain −20 dB input match in the filter passband. The designed filter is 26 mm long and realizable with double-coupled lines that require a minimum spacing of 0.079 mm at the coupler-filter center.

The filter can be accommodated on a smaller substrate by folding the coupled lines. The folded filter is illustrated in Figure 7.14. The number of horizontal sections should be an odd number to preserve symmetricity in order not to disturb the phase quadrature in the filter passband. The folded structure shown in Figure 7.14(a) is relatively straightforward compared to a folded structure with a large number of bends; however, it may cause a larger distortion on the filter performance unless the additional length introduced by the bends is taken into account. The computed performance is shown in Figure 7.14(b). A −3 dB bandwidth extends from 14 to 18 GHz with less than 0.1 dB ripple. The rejection level is low, with a minimum value of 14 dB in the frequency range of the filter. Neither the bandpass coupling nor the out-of-band coupling of this filter has been optimized. The same is true with all the other coupler-filter circuits presented so far.

It is not difficult to predict that the lossses of nonuniform line filters are high compared to waveguide filters but the cost reduction is at least 100-fold. The waveguide filters or any other filter design method I know do not have the same ease in design and frequency range of applicability as offered by nonuniform coupled-line bandpass filters.

7.6.3 X-Band Bandpass Filters on Alumina Substrate with $\epsilon_r = 9.9$ and $h = 0.635$ mm

Two more examples are given for bandpass filters utilizing the principle of reflection at the input. The first filter is designed to operate in the 8–12 GHz band. The filter length is 33 mm, which gives the desired bandpass skirt sharpness; the coupling imbalance is less than ±0.2 dB, which gives a better than −30 dB input match. The computed performance is shown in Figure 7.15(a); the passband ripple is less than 0.05 dB.

The second design is for a narrower bandwidth, 9–11 GHz, whose computed performance is given in Figure 7.15(b). The effect of a 5.4-mm-long 50 Ω microstrip line connected to the direct port of the coupler-filter on the performance is computed and shown in Figure 7.15(c). We see that it is possible to realize very high-performance filters with minimum effort spent on the minimization of the out-of-band coupling lobes. Because the phase introduced by the 50 Ω microstrip line is linear, the passband is slightly distorted; out-of-band couplings at 5 and 15 GHz regions are eliminated.

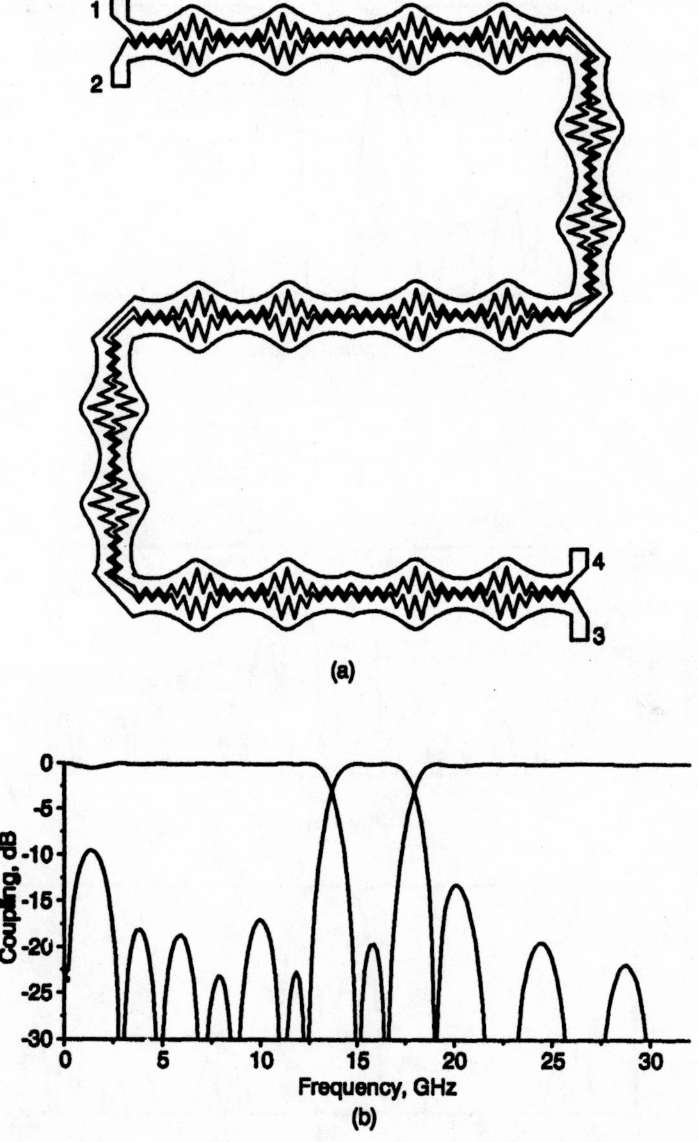

Figure 7.14 (a) Folded filter with minimum bends. (b) Computed performance of a K_u-band reflection-type filter.

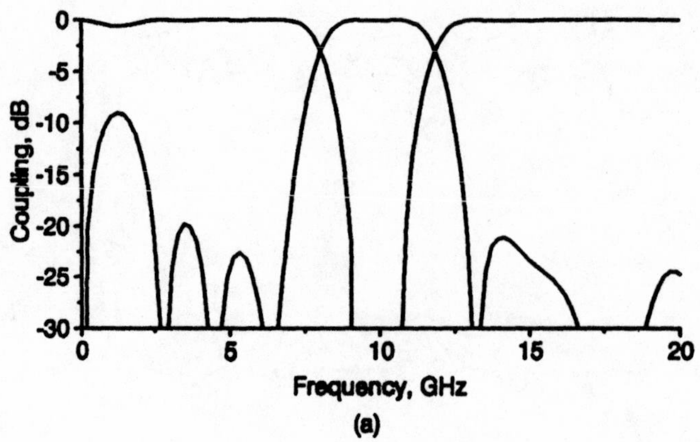

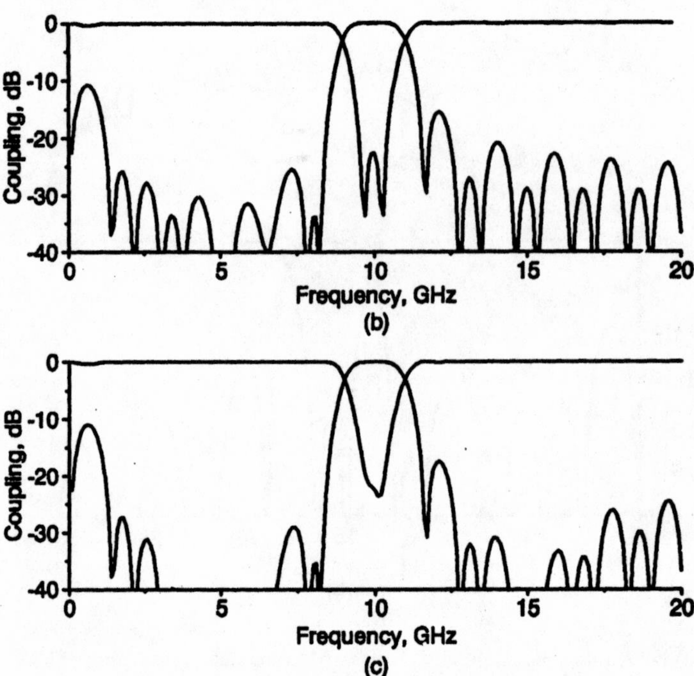

Figure 7.15 Computed performance for *X*-band reflection-type filters: (a) 8–12 GHz filter with l = 33 mm, (b) 9–11 GHz filter with l = 50 mm, and (c) 9–11 GHz filter with l = 50 mm and a 5.4-mm line added to the direct port.

Chapter 8
Periodic Nonuniform Line Coupler-Filters and Multiplexers

8.1 INTRODUCTION

In wideband receivers, channelization of the input is required to perform several functions, such as the minimization of intermodulation products, baseband conversion for detection, and analysis and sorting for radar, communication, and surveillance systems. The multiplexer is the key component for channelized receivers. The design of low-cost, small-size, lightweight channelizers is a difficult task. Waveguide components, especially at millimeter waves, require precision-machined parts, causing an astronomical increase in production cost.

The most commonly used wideband channelizer configuration requires eight bandpass filters and eight -3 dB directional couplers. Wideband channelizers at K_a-band using waveguide components [1], using suspended substrate [2], and at *W*-band using monolithic technology [3] have been reported for contiguous channelization of a wideband input signal using a similar configuration. A 2–18 GHz wideband quadruplexer using four tandem connected 2–18 GHz, -3 dB couplers and phase-shifting elements has also been reported [4]. Highly repeatable, lightweight, low-cost, and small-size multiplexers with added design flexibility and using just four bandpass coupler-filters will be described in this chapter.

Periodic coupler-filters eliminate the need for multioctave feed networks for multifrequency antennas. The required amplitude of excitation for a single multifrequency antenna element (the antenna element may be one of several antenna elements of an array) can be obtained by adjusting the coupling levels of a periodic-type directional coupler. Because the coupling is of bandpass type the need for additional filtering is eliminated, resulting in superior performance with significant cost reduction.

8.2 DESIGN PRINCIPLES

The general design procedure presented in Chapter 4 combined with the proposed modifications for bandpass directional couplers given in Chapter 6 can be used in the design of periodic coupler-filters. In this case an additional coupling channel (or channels) is specified in a given frequency range. With two coupling channels the reflection coefficient distribution function is modified as follows:

$$\begin{aligned} p(x) = -\frac{2}{\pi v}\Bigg\{ & \int_0^{\omega_d} \sin(2\omega x/v)\tanh^{-1}[C_d(\omega)]\,d\omega \\ & + \int_{\omega_d}^{\omega_{B_1}} \sin(2\omega x/v)\tanh^{-1}[C_1(\omega)]\,d\omega \\ & + \int_{\omega_{B_1}}^{\omega_{B_2}} \sin(2\omega x/v)\tanh^{-1}[C_{B_1}(\omega)]\,d\omega \\ & + \int_{\omega_{B_2}}^{\omega_{B_3}} \sin(2\omega x/v)\tanh^{-1}[C_2(\omega)]\,d\omega \\ & + \int_{\omega_{B_3}}^{\omega_{B_4}} \sin(2\omega x/v)\tanh^{-1}[C_{B_2}(\omega)]\,d\omega \\ & + \int_{\omega_{B_4}}^{2\omega_c} \sin(2\omega x/v)\tanh^{-1}[C_3(\omega)]\,d\omega \Bigg\} \end{aligned} \tag{8.1}$$

where $C_{B_1}(\omega)$ and $C_{B_2}(\omega)$ are the specified couplings in the desired channel bandwidths. $C_1(\omega)$, $C_2(\omega)$, and $C_3(\omega)$ correspond to out-of-band regions and can be set to be equal to zero. Again, a dummy coupling channel, $C_d(\omega)$, is used for physical realizability. The choice of $2\omega_c$ depends on the application, and it may be set to be equal to ω_{B4}.

A typical performance obtainable by using (8.1) is given in Figure 8.1(a) for −3 dB bandpass coupling in the 12–20 GHz and 28–36 GHz channels. The continuous coupling coefficient for this design is given in Figure 8.1(b). A close inspection of $k(x)$ reveals that the coupling coefficients at the vicinity of the center are not realizable with double-coupled lines on alumina substrate. It is interesting to note that the first channel is at microwave frequencies, where the second channel extends into the millimeter wave region. Therefore, the choice of substrate thickness becomes very critical.

An alumina substrate with a standard 0.635 mm thickness is good enough for the first channel but surface waves will significantly deteriorate the performance of the second channel. The use of an alumina substrate with a smaller thickness, say, 0.381 mm, will introduce difficulties for the use of an interdigitated section due to reduced

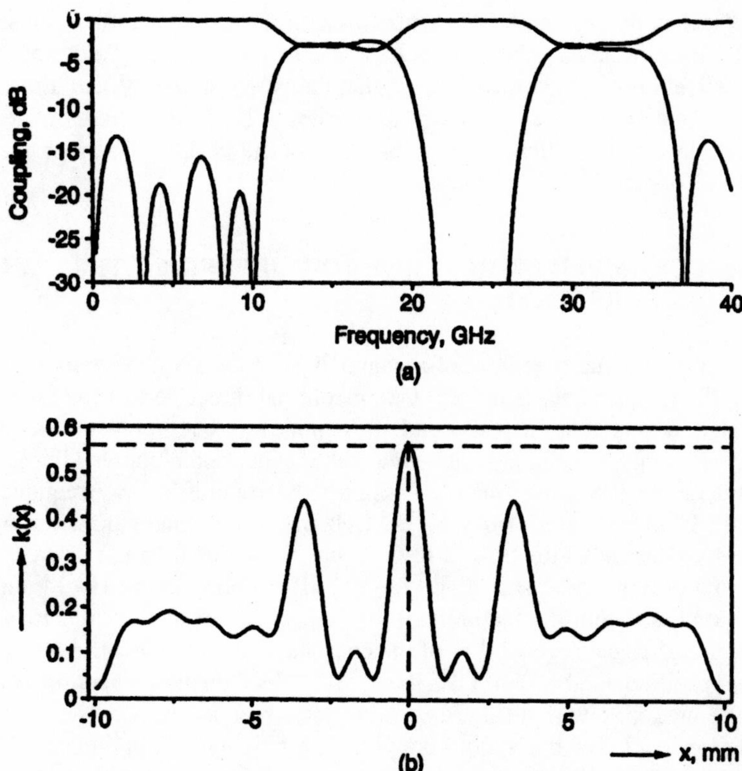

Figure 8.1 Typical periodic coupler-filter: (a) performance and (b) continuous-coupling coefficient.

physical dimensions required for this section. With the indicated $k(0) = 0.56$, the required width is about 0.048 mm with $N = 4$ on a 0.381-mm-thick alumina substrate. This is just about the practically realizable limit with wet etching techniques; it is easier to realize this width with *e*-beam writing. The continuous-coupling coefficient can be realized using double-coupled lines on 0.2-mm-thick GaAs MMIC technology. However, in this case, the longer length may constitute a problem. Folding the structure should be possible for space optimization. Some of the key design features of periodic coupler-filters (using hybrid MIC or MMIC technology) are highlighted with the preceding example. It is also worth noting that the two coupling-coefficient lobe amplitudes on either side of the center also depend on the channel bandwidths, specified channel-coupling amplitudes, the periodic coupler-filter length, and the number of channels in a given frequency range. High coupling amplitudes at these regions should be avoided as they may otherwise require interdigitated sections, adding to the complexity of the design.

A typical periodic coupler-filter circuit with a smaller length is illustrated in Figure 8.2. This circuit was realized on 25.4 × 25.4 × 0.635 mm^3 alumina substrate with $\epsilon_r = 9.9$ and $h = 0.635$ mm. We see that the variation in physical dimensions is rather smooth, thereby avoiding sharp taper rates. This is an inhererent feature of periodic coupler-filters, which can also be used in the design of bandpass filters as suggested in Chapter 7.

8.3 DESIGN EXAMPLES FOR PERIODIC COUPLER-FILTERS ON ALUMINA SUBSTRATE

Tight couplings are usually required for channelization of a wideband input signal. In such cases, the first channel is usually taken from the direct port of the coupler-filter because a tight coupling at the lower end of the microwave region would require high coupling coefficients, which may not be realizable with double-coupled lines. This is a direct result of the increased fractional channel bandwidth at low frequencies. For example, a 4-GHz bandwidth from 34–38 GHz has a fractional bandwidth of 11.1% whereas a 4-GHz bandwidth from 2–6 GHz has a fractional bandwidth of 100%.

The first design example is given for -3 dB coupling in the 10–14 and 18–22 GHz bands on 0.635-mm-thick alumina substrate. The computed results are shown in Figure 8.3. For physical realizability of the continuous coupling coefficient, a dummy channel is specified in the 0–1 GHz band. Its effect on the reflection coefficient distribution function, given in Figure 8.3(a), can be seen as an increase in $p(x)$ at the vicinity of $x = -l/2$. The continuous coupling coefficient given in Figure 8.3(b) has a maximum value of 0.438 at $x = 0$. We see that $k(x)$ does not have a large number of

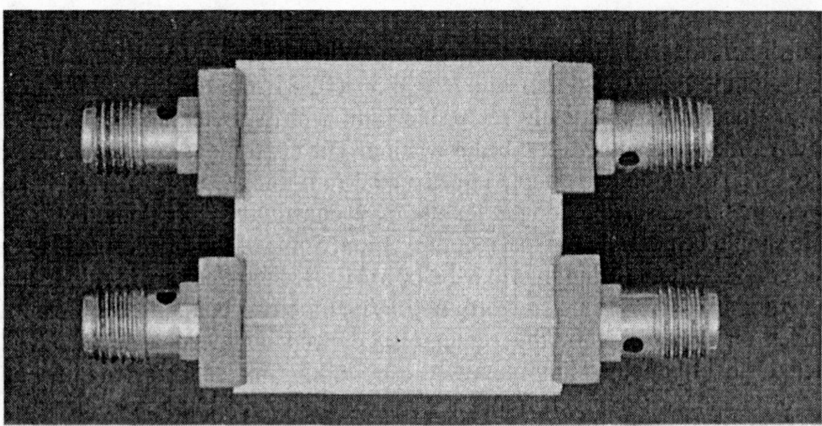

Figure 8.2 Periodic coupler-filter.

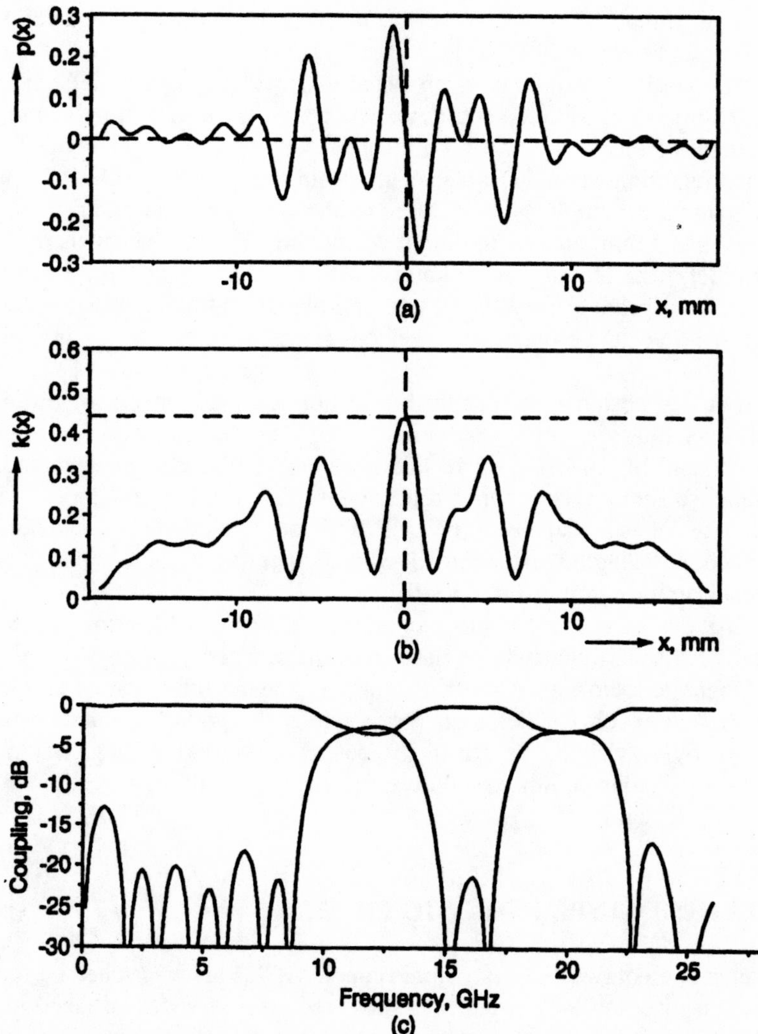

Figure 8.3 Computed results for wideband matched periodic coupler-filter on alumina: (a) reflection coefficient distribution function, (b) continuous-coupling coefficient, and (c) performance.

minima and maxima. The computed performance is shown in Figure 8.3(c). Out-of-band coupling is below −20 dB.

The next design example is given for an ultrawideband periodic coupler-filter design on 0.5-mm-thick alumina substrate. The computed results are given in Figure 8.4. The length of this design is by no means optimum and it also depends on the performance requirements using a selected substrate technology. On 0.2-mm-thick GaAs substrate the 30-mm length is reduced to about 26 mm, which can be accommodated in less than 5 mm^2 area by folding the structure. With two channels (6–10 GHz and 14–18 GHz) the length can be halved with a reduction in $k(0)$. The coupling balance in the respective channels is not optimized. Optimum performance for a specific application can be obtained by selective optimization (adding dummy channels).

The last two design examples in this section are given to illustrate the use of periodic coupler-filters in narrowband applications. The first narrowband design is for a 6–8, 10–12, and 14–16 GHz −3 dB periodic coupler-filter designed on 0.635-mm-thick alumina substrate. The computed results are given in Figure 8.5. The value of $k(0)$ is 0.463, which is realizable with $N = 2$. The coupling-coefficient variation along most of the coupler length is rather small, which suggests that there is plenty of room for additional optimization to reduce $k(0)$.

The final design example in this section is given for −3 dB coupling in the 8–9, 10–11, and 12–13 GHz channels. We see from the computed results given in Figure 8.6 that a longer periodic coupler-filter is required to adequately separate the channels. Naturally, $k(0)$ is much smaller than the previous design. It is again possible to optimize $k(0)$ by varying the length of the design. Because the frequency range of interest is greatly reduced, other modifications of $p(x)$ should be investigated for an optimum design length.

8.4 REFLECTION-TYPE PERIODIC FILTERS

Tandem connection of two −3 dB periodic coupler-filters gives matched performance with minimum insertion loss in the respective channels. A matched performance is required because the direct port may be used to obtain the other channels or for other purposes, such as a channel-dropping filter. However, when only the coupled channels are required for an specific application, any mismatch at the input outside the channels would not be a problem. Therefore, it would be sufficient to use −3 dB periodic coupler-filters with open-circuited coupled and direct ports, where the periodic filter output is obtained from the isolated port. The amplitude of the reflected signal in the design channels depends on the coupling balance and phase quadrature in those channels. Similarly, the ripple amplitude depends on these design requirements.

The computed performance of a reflection-type periodic filter looks exactly the same as that of a tandem circuit given by the previous design examples. The direct

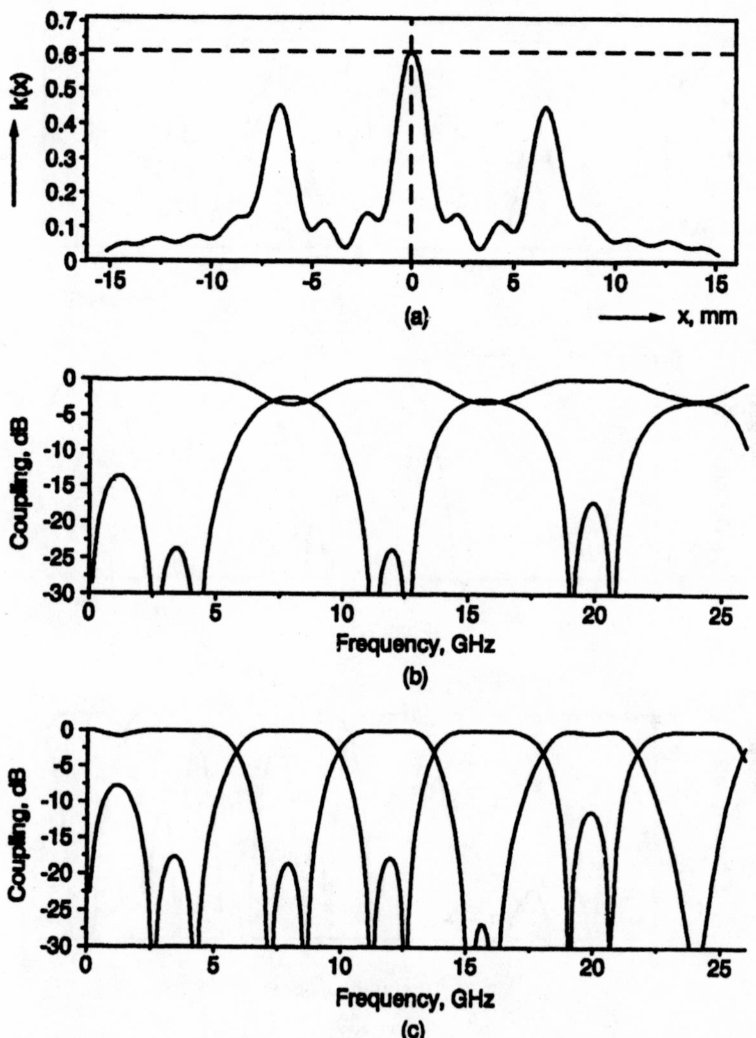

Figure 8.4 Computed results for an ultrawideband matched periodic coupler-filter on alumina: (a) continuous-coupling coefficient, (b) performance of periodic coupler, and (c) performance of periodic filter.

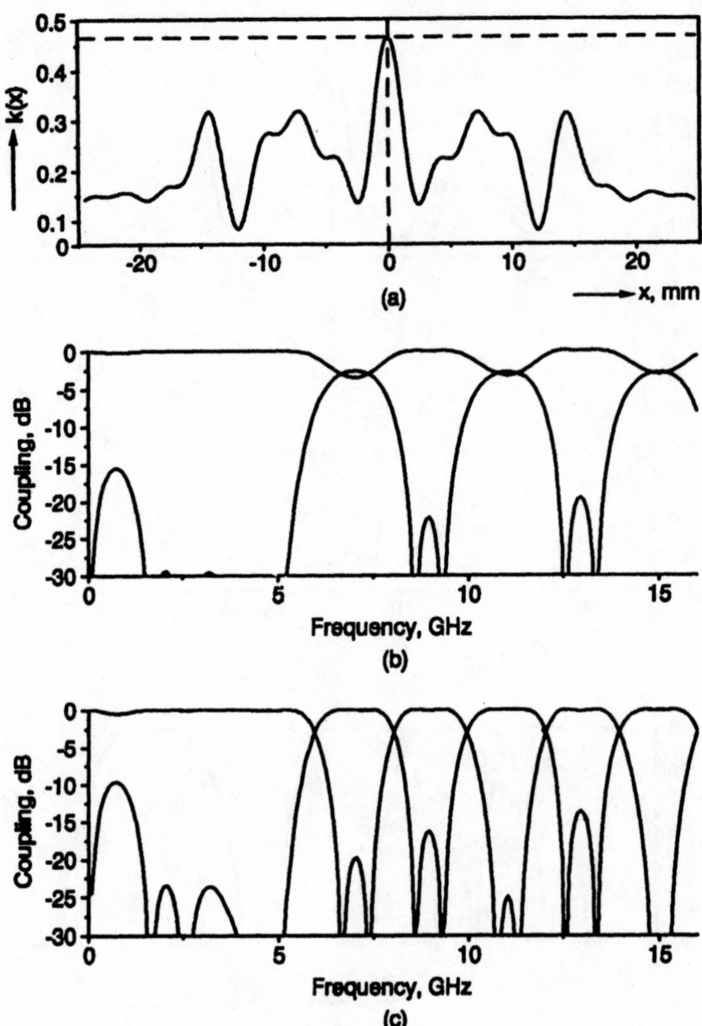

Figure 8.5 Computed results for a matched periodic coupler-filter with 2 GHz channel bandwidths: (a) continuous-coupling coefficient, (b) performance of periodic coupler, and (c) performance of periodic filter.

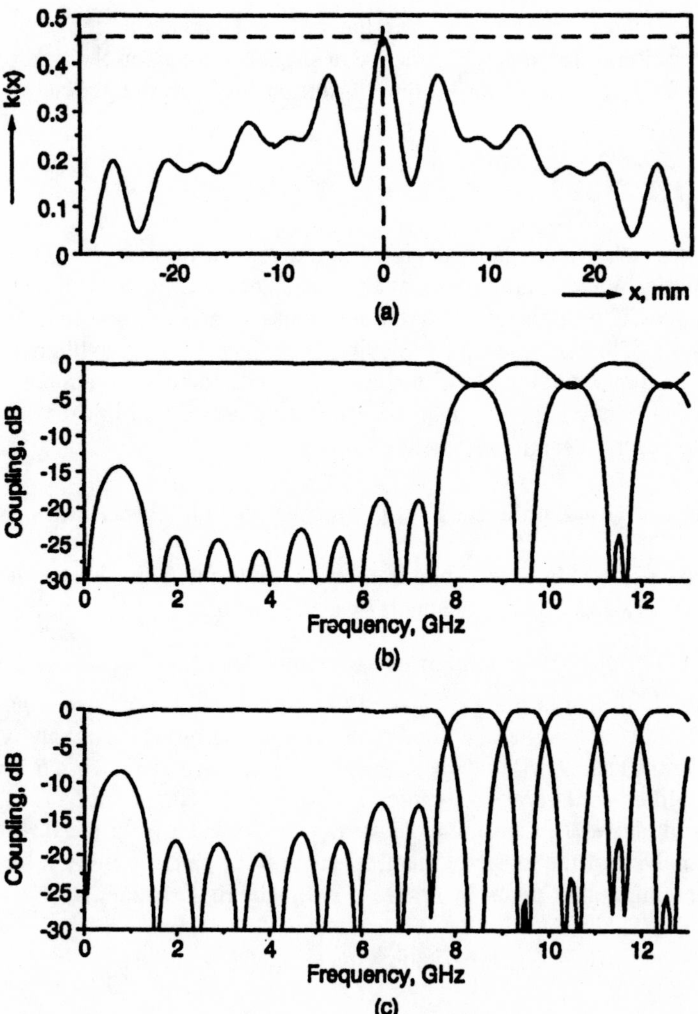

Figure 8.6 Computed results for a matched periodic coupler-filter with 1 GHz channel bandwidths: (a) continuous-coupling coefficient, (b) performance of periodic coupler, and (c) performance of periodic filter.

signal of the tandem circuit becomes the reflected signal for the single periodic coupler-filter circuit. However, in practice, a slight difference in their performance is expected mainly because of the channel insertion loss and the geometry of tandem configuration.

8.5 MULTIPLEXERS

We must separate a wideband input signal into a number of individual channels. With periodic coupler-filters, multiplexing of an input signal becomes straightforward. For a moderate bandwidth duplexer (two-channel multiplexer), a single matched coupler-filter can do the job. For wideband duplexing, two matched coupler-filters in tandem is adequate. The number of coupler-filters increases with increasing number of channels and increasing bandwidth. A quadruplexer (four-channel multiplexer) is chosen to demonstrate several design approaches using coupler-filters.

8.5.1 Design of Quadruplexers Using Almost 0 dB Matched Bandpass Filters

The four-channel multiplexer configuration is shown in Figure 8.7(a). A wideband input signal is divided into four channels by using four almost 0 dB coupler-filters. The isolated ports and the direct port of the last coupler-filter are terminated. With a clean input signal the number of coupler-filters can be reduced to three. This simplified configuration is shown in Figure 8.7(b). The first channel is obtained from the direct port of the third coupler-filter. Isolated ports are again terminated. If there is accompanied noise below channel B_1 that needs to be adequately separated from B_1, we can still use the simplified configuration by interchanging channels B_4 and B_1. If noise occurs below B_1 and above B_4, then either B_2 or B_3 can be taken from the direct port.

The analysis of the matched quadruplexer can be carried out by referring to the simplified configuration given in Figure 8.7(b). The first channel B_1 is obtained by

$$B_1 = C_{B_1}(\omega) = \tanh\left\{\int_0^{l_1} \sin(2\omega x/v)p_1(x)\,\mathrm{d}x\right\} \tag{8.2}$$

where l_1 is the length of the filter and $p_1(x)$ is its modified reflection coefficient distribution function.

$D_1(\omega)$ can be obtained from $C_{B_1}(\omega)$ as

$$D_1(\omega) = \sqrt{1 - C_{B_1}^2(\omega)} \tag{8.3}$$

or

$$D_1(\omega) = \frac{1}{\cosh\left\{\int_0^{l_1} \sin(2\omega x/v)p_1(x)\,\mathrm{d}x\right\}} \tag{8.4}$$

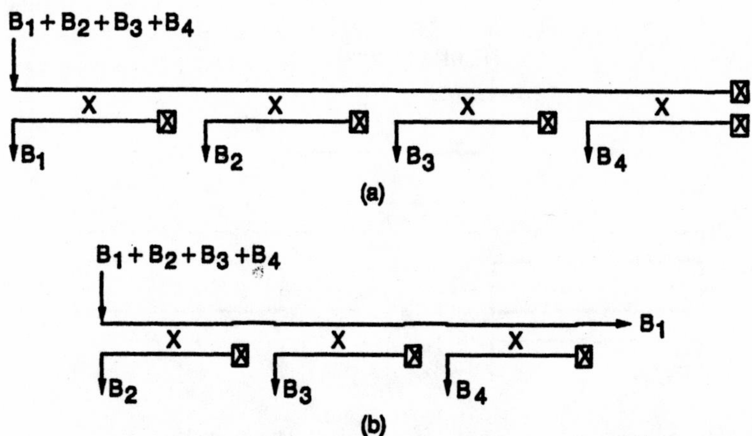

Figure 8.7 Quadruplexer circuits using almost 0 dB matched filters: (a) general circuit and (b) simplified circuit.

The second channel is then given by

$$B_2 = C_{B_2}(\omega) = D_1(\omega)\tanh\left\{\int_0^{l_2} \sin(2\omega x/v)p_2(x)\,\mathrm{d}x\right\} \tag{8.5}$$

where l_2 is the length of the filter and $p_2(x)$ is its modified reflection coefficient distribution function. The direct signal of this coupler-filter is

$$D_1(\omega) = \sqrt{D_1^2(\omega) - C_{B_2}^2(\omega)} \tag{8.6}$$

For the last section we have

$$B_3 = C_{B_3}(\omega) = D_2(\omega)\tanh\left\{\int_0^{l_3} \sin(2\omega x/v)p_3(x)\,\mathrm{d}x\right\} \tag{8.7}$$

$$B_4 = D_3 = \sqrt{D_2^2(\omega) - C_{B_3}^2(\omega)} \tag{8.8}$$

8.5.2 Design of Quadruplexers Using −3 dB Matched Coupler-Filters

A possible quadruplexer configuration using −3 dB matched coupler-filters is given in Figure 8.8. The first tandem filter is a periodic type separating the input signal into B_1, B_3 (direct), and B_2, B_4 (coupled) signals. The individual channels are obtained by tandem-connected filters having passbands B_3 and B_4. The isolated ports of the tandem filters are terminated. Assuming identical coupler-filters for each tandem filter with

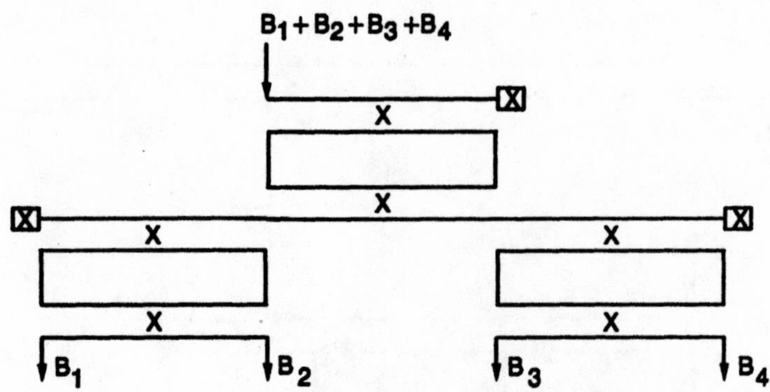

Figure 8.8 Quadruplexer circuit using −3 dB coupler-filters in tandem.

coupling functions $C_1(\omega)$, $C_2(\omega)$, and $C_3(\omega)$, respectively, the channel responses are obtained by

$$B_1 = [1 - 2C_1^2(\omega)][1 - 2C_2^2(\omega)] \tag{8.9}$$

$$B_2 = [2C_1(\omega)C_3(\omega)\sqrt{1 - C_1^2(\omega)}][1 - 2C_3^2(\omega)] \tag{8.10}$$

$$B_3 = [1 - 2C_1^2(\omega)][2C_1(\omega)C_2(\omega)\sqrt{1 - C_2^2(\omega)}] \tag{8.11}$$

$$B_4 = [4C_1(\omega)C_2(\omega)\sqrt{[1 - C_1^2(\omega)][1 - C_2^2(\omega)]} \tag{8.12}$$

with

$$C_1(\omega) = \tanh\left\{\int_0^{l_1} \sin(2\omega x/v)p_1(x)\,dx\right\} \tag{8.13}$$

$$C_2(\omega) = \tanh\left\{\int_0^{l_2} \sin(2\omega x/v)p_2(x)\,dx\right\} \tag{8.14}$$

$$C_3(\omega) = \tanh\left\{\int_0^{l_3} \sin(2\omega x/v)p_3(x)\,dx\right\} \tag{8.15}$$

8.5.3 Design of Quadruplexers Using −3 dB Reflection-Type Coupler-Filters

Both filtering and channelization of a wideband input signal is also possible by using −3 dB reflection-type coupler-filters. The quadruplexer realized by using four such circuits is shown in Figure 8.9. Delay lines d_1, d_2, and d_3 are connected to the coupled

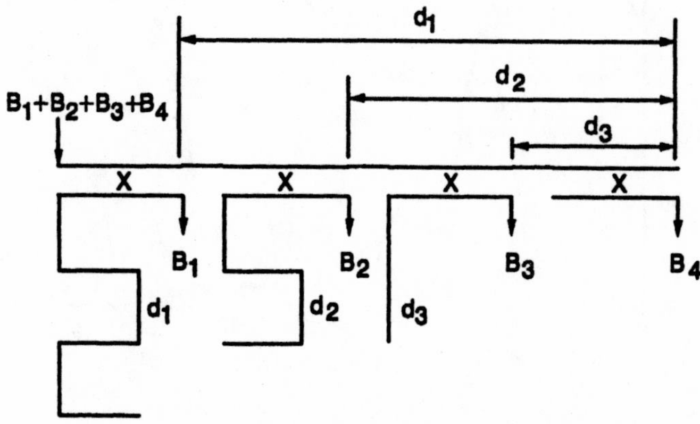

Figure 8.9 Quadruplexer circuit using −3 dB reflection-type coupler-filters.

ports of the first, second, and third coupler-filters, respectively. The desired channels B_1, B_2, B_3, and B_4 are obtained as indicated. No termination is required for this multiplexer. The circuit may be bent or shaped as desired. Delay lines connected to the coupled ports can be designed in any way and any shape as long as they provide the required delay to maintain phase quadrature in the channel bandwidths.

The required channels are obtained by

$$B_1 = 2C_1(\omega)\sqrt{1 - C_1^2(\omega)}[1 - C_2^2(\omega)][1 - C_3^2(\omega)][1 - C_4^2(\omega)] \tag{8.16}$$

$$B_2 = 2C_2(\omega)\sqrt{[1 - C_1^2(\omega)][1 - C_2^2(\omega)]}[1 - C_3^2(\omega)][1 - C_4^2(\omega)] \tag{8.17}$$

$$B_3 = 2C_3(\omega)\sqrt{[1 - C_1^2(\omega)][1 - C_2^2(\omega)][1 - C_3^2(\omega)]}[1 - C_4^2(\omega)] \tag{8.18}$$

$$B_4 = 2C_4(\omega)\sqrt{[1 - C_1^2(\omega)][1 - C_2^2(\omega)][1 - C_3^2(\omega)][1 - C_4^2(\omega)]} \tag{8.19}$$

The design of this type of quadruplexer is illustrated by an example at K_a-band on 0.2-mm-thick GaAs substrate. We must separate a 26–42 GHz input signal into 4-GHz-wide channels. Because the widest fractional bandwidth is only 14.3%, the coupler-filters can be realized with double-coupled lines. The performance of the quadruplexer is obtained by designing the four individual coupler-filters and then computing the individual channel performances by using eqs. (8.16) through (8.19). The computed performance is given in Figure 8.10. Signal leaking into the adjacent channels is kept below -20 dB. Attenuation at 1 GHz away from channel cross-over frequencies is 20 dB.

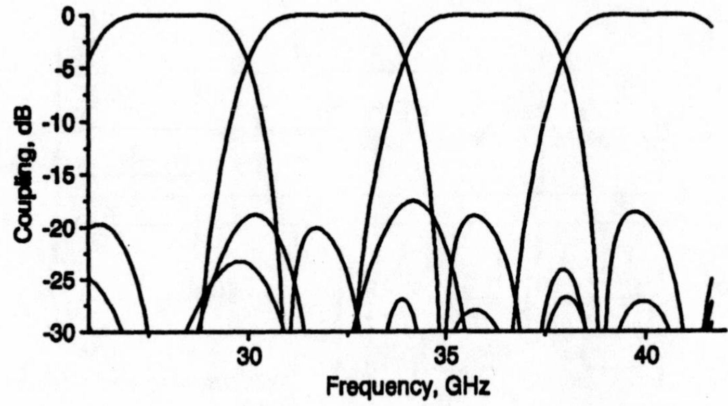

Figure 8.10 Computed performance for a K_a-band quadruplexer on GaAs.

REFERENCES

[1] Breuer, K. D., and N. Worontzoff, "A Low Cost Multiplexer for Channelized Receiver Front Ends at Millimeter Waves," *IEEE MTT-S Int. Microwave Symp. Digest,* pp. 150–152.

[2] Rubin, D., "Millimeter Wave Hybrid Coupled Reflection Amplifiers and Multiplexers," *IEEE Trans. on Microwave Theory and Tech.,* Vol. MTT-30, 1982, pp. 2156–2162.

[3] Lan, G. L., J. C. Chen, C. K. Pao, and M. I. Herman, "A W-Band Channelized Monolithic Receiver," *IEEE Microwave and Millimeter-Wave Monolithic Circuits Symp. Digest,* 1989, pp. 95–99.

[4] Uysal, S., A. H. Aghvami, and S. A. Mohamed, "Microstrip Quadruplexer Using −3 dB Hybrids," *Proceedings of the 19th European Microwave Conference,* 1989, pp. 905–910.

Chapter 9
Codirectional Nonuniform Line Directional Couplers

9.1 INTRODUCTION

Until now we have studied nonuniform line backward-wave directional couplers. In this chapter we are going to change the direction of the coupled signal. In microstrip couplers there is a finite signal appearing at all four ports: S_{11}, S_{22}, S_{31}, and S_{41}. With quarter-wavelength-long uniform coupled lines S_{21} is larger than S_{31} at the design center frequency; hence the need for modification to maximize the signal appearing at the so-called coupled port. Once S_{31} is eliminated (minimized) S_{21} becomes a well-behaved function, which enables the designer to use cascaded sections for wideband performance. However, in backward-wave couplers the amount of coupled signal is dictated by the gap between the two lines. In the case of forward-wave (codirectional, because the coupled signal is in the same direction with the direct signal) couplers we have an entirely different coupling mechanism, which is dictated by the difference of the even- and odd-mode phase velocities. But what about the spacing between the two lines? If the two lines are separated by an infinite(!) gap ($s \approx 5$ mm), there is no coupling between them. Hence, there is no even or odd mode; there is only one velocity. This means that the even- and odd-mode phase velocities are a function of the gap between the two lines. When s decreases, the phase velocity difference increases up to a certain value of s; at this value of s, the phase velocity difference starts to decrease because we have a second limit, which is the limit when s approaches zero. In the limit $s \rightarrow 0$ we no longer have coupled lines: we have a single line with $w = 0$.

Because a microstrip is dispersive, phase velocity at $k = 0$ ($s = \infty$) will be lowered but the upper limit on the coupled lines is fixed, as $w = 0$ with $s = 0$. Now we can draw the final picture, which is given in Figure 9.1.

The phase velocity at $k = 0$ is the velocity of a single microstrip with width w, given by

$$v_m(f) = \frac{c}{\sqrt{\epsilon_{reff}(f)}} \tag{9.1}$$

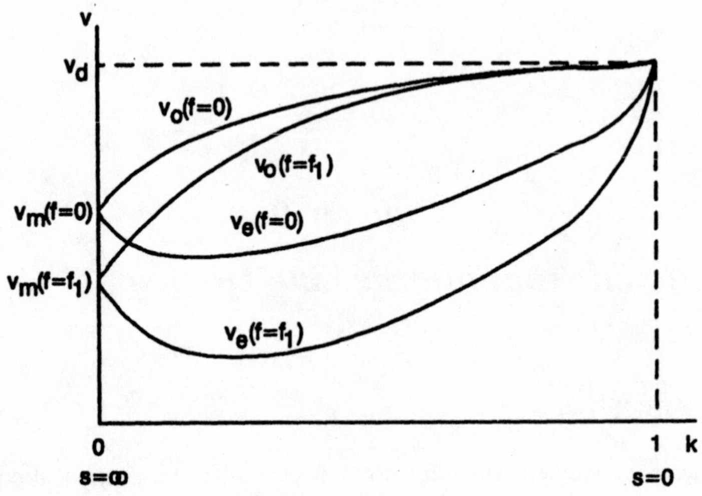

Figure 9.1 Variation of even- and odd-mode phase velocities of coupled lines with coupling coefficient and frequency.

where c is the speed of light and $\epsilon_{reff}(f)$ is the frequency-dependent effective dielectric constant.

The phase velocity at $k = 1$ is the velocity at the dielectric air interface ($w = 0$), given by

$$v_d = \frac{c}{\sqrt{1 + \epsilon_r/2}} \tag{9.2}$$

where ϵ_r is the relative dielectric constant.

9.2 ANALYSIS OF CODIRECTIONAL COUPLERS

The scattering parameters in inhomogeneous media can be expressed in terms of coupled and direct signals as follows:

$$S_{11} = \frac{1}{2}[C_e(\omega) + C_o(\omega)] \tag{9.3}$$

$$S_{21} = \frac{1}{2}[C_e(\omega) - C_o(\omega)] \tag{9.4}$$

$$S_{31} = \frac{1}{2}[D_e(\omega) - D_o(\omega)] \tag{9.5}$$

$$S_{41} = \frac{1}{2}[D_e(\omega) + D_o(\omega)] \tag{9.6}$$

To determine the four S-parameters we need to solve the nonlinear differential equation of a nonuniform line for both the even and odd modes and obtain the final solution by the superposition of these two modes as indicated by the preceding equations. Under matched condition, the solution of the nonlinear differential equation can be rewritten for the coupled and direct signals of the two modes as follows:

$$C_e(\omega) = \tanh(G_e)[\cos(2\omega l/v_e) + \mathrm{j}\sin(2\omega l/v_e)] \tag{9.7}$$

$$C_o(\omega) = \tanh(G_o)[\cos(2\omega l/v_o) + \mathrm{j}\sin(2\omega l/v_o)] \tag{9.8}$$

$$D_e(\omega) = \frac{1}{\cosh(G_e)}[\cos(2\omega l/v_e) + \mathrm{j}\sin(2\omega l/v_e)] \tag{9.9}$$

$$D_o(\omega) = \frac{1}{\cosh(G_o)}[\cos(2\omega l/v_o) + \mathrm{j}\sin(2\omega l/v_o)] \tag{9.10}$$

where

$$G_e = \int_0^l \sin(2\omega x/v_e)p_e(x)\,\mathrm{d}x \tag{9.11}$$

$$G_o = \int_0^l \sin(2\omega x/v_o)p_o(x)\,\mathrm{d}x \tag{9.12}$$

S-parameters are then obtained by

$$S_{11} = \frac{1}{2}\sqrt{\tanh^2(G_e) + \tanh^2(G_o) + 2\tanh(G_e)\tanh(G_o)\cos\left[2\omega l\left(\frac{1}{v_e} - \frac{1}{v_o}\right)\right]} \tag{9.13}$$

$$S_{21} = \frac{1}{2}\sqrt{\tanh^2(G_e) + \tanh^2(G_o) - 2\tanh(G_e)\tanh(G_o\cos\left[2\omega l\left(\frac{1}{v_e} - \frac{1}{v_o}\right)\right]} \tag{9.14}$$

$$S_{31} = \frac{1}{2}\sqrt{\frac{1}{\cosh^2(G_e)} + \frac{1}{\cosh^2(G_o)} - \frac{2\cos[2\omega l(1/v_e - 1/v_o)]}{\cosh(G_e)\cosh(G_o)}} \tag{9.15}$$

$$S_{41} = \frac{1}{2}\sqrt{\frac{1}{\cosh^2(G_e)} + \frac{1}{\cosh^2(G_o)} + \frac{2\cos[2\omega l(1/v_e - 1/v_o)]}{\cosh(G_e)\cosh(G_o)}} \tag{9.16}$$

where G_e and G_o are the backward-wave coupling for the even and odd modes, respectively. Because S_{31} is considered as the main coupled signal, to maintain perfect isolation (i.e., infinite directivity), we require $G_e = G_o = 0$, which in turn gives $S_{11} = S_{21} = 0$. In this case, S_{31} and S_{41} can be simplified to

$$S_{31} = \sin\left[\omega l\left(\frac{1}{v_e} - \frac{1}{v_o}\right)\right] \tag{9.17}$$

$$S_{41} = \cos\left[\omega l\left(\frac{1}{v_e} - \frac{1}{v_o}\right)\right] \tag{9.18}$$

From these equations we can deduce that

$$|\text{angle}(S_{31}) - \text{angle}(S_{41})| = \frac{\pi}{2} \tag{9.19}$$

The sign of the phase quadrature depends on the magnitude of v_e and v_o. As explained in the introduction section we need some coupling to have even and odd modes. This means that we cannot immediately use the simplified equations. We have to use the general equations. Because we need high directivity we must have G_e and G_o negligibly small. Assume, for the time being, that -20 dB isolation is satisfactory. This requires $G_e \approx 0.1$ and $G_o \approx -0.1$. The difference between even- and odd-mode phase velocities will be small, which means that a very long coupler would be required for a high S_{31} at microwave frequencies. If isolation is further reduced to -30 dB a much longer coupler would be required to achieve high forward-wave coupling. The question then arises whether it is possible to maximize S_{31} and at the same time reduce S_{21} without increasing the coupler length. Referring to Figure 9.1 we can see that we need coupling coefficients larger than 0.1 to use a larger phase velocity difference of the two modes. A very effective technique can provide the required solution for codirectional couplers. The bandpass coupling technique, which we studied in Chapter 6, can also be applied to nonuniform line codirectional couplers. A dummy channel or channels can be specified to obtain a physical coupler with $k(0)$ maintained at any value that would give the desired minimum spacing. Because backward-wave coupling is far away from the design bandwidth, very high directivities can be achieved with couplers having reasonable lengths.

9.3 DESIGN CURVES FOR CODIRECTIONAL COUPLERS

In uniform microstrip couplers, phase velocities are functions of both frequency and the coupling coefficient. In nonuniform microstrip couplers, because coupling coefficient becomes a function of x, phase velocities can be expressed as functions of frequency and distance along the coupler. That is, we now have $v_e(\omega, x)$ and $v_o(\omega, x)$. Therefore the analysis given in the previous section is valid only with constant phase velocities. If we use $v_e(\omega, x)$ and $v_o(\omega, x)$ in G_e and G_o, we cannot obtain $p_e(x)$ and $p_o(x)$ by a simple inverse Fourier transform. Even if we neglect the variation of phase velocities with respect to x we still have a strong frequency dependence for the range of coupling coefficients in which we are most interested. However, we can still use an iterative solution to obtain the four S-parameters. This is indeed possible as we can use any $p(x)$ (as long as it gives a realizable $k(x)$) to compute the backward-wave coupling. Because we are not interested in maintaining a nominal coupling or coupling balance for the backward-wave coupler, its general shape, whether distorted or not, is not significant. This means that by using an average phase velocity, for the even and odd modes, we calculate $p_e(x)$ and $p_o(x)$ and then G_e and G_o. However, when we compute the S-parameters, we use $v_e(\omega)$ and $v_o(\omega)$ instead of v_e and v_o. If we want to include x dependence, l/v_e and l/v_o are replaced by

$$\int_0^l \frac{dx}{v_e(\omega, x)} \qquad \int_0^l \frac{dx}{v_o(\omega, x)} \tag{9.20}$$

Since we have a satisfactory, realizable $k(x)$ we can use it to obtain $v_e(\omega, x)$ and $v_o(\omega, x)$ by using inverse cubic spline interpolation. Alternatively, the design curves (phase velocities versus coupling coefficient), which will be given in the following subsections for some important substrates, can be used to obtain the phase velocities. In the following computations, impedance level is taken as 50 Ω.

9.3.1 Even- and Odd-Mode Phase Velocities versus Coupling Coefficient for Alumina Substrate with $\epsilon_r = 9.9$ and $h = 0.635$ mm, 0.25 mm, and 0.1 mm

The design information given in Section 3.2 or the design steps 1–17 (excluding phase velocity compensation) given in Chapter 4 can be used to compute the even- and odd-mode phase velocities for uniform coupled lines. The desired $v_e(\omega, x)$ and $v_o(\omega, x)$ are then obtained by cubic spline interpolation.

The computed even- and odd-mode phase velocities for alumina substrate with $h = 0.635$ mm, 0.25 mm, and 0.1 mm are given in Figures 9.2, 9,3, and 9.4, respectively. The phase velocities for $h = 0.635$ are computed from $f = 5$–20 GHz in steps of 5 GHz and for $h = 0.25$ and 0.1 mm from $f = 25$–55 GHz and $f = 65$–95 GHz, respectively, in steps of 10 GHz. As it can be seen from the computed results, the thicker the substrate, the larger is the dispersion. However, for thicker substrates variation in even-mode phase velocity with respect to x is very small for $k < 0.4$.

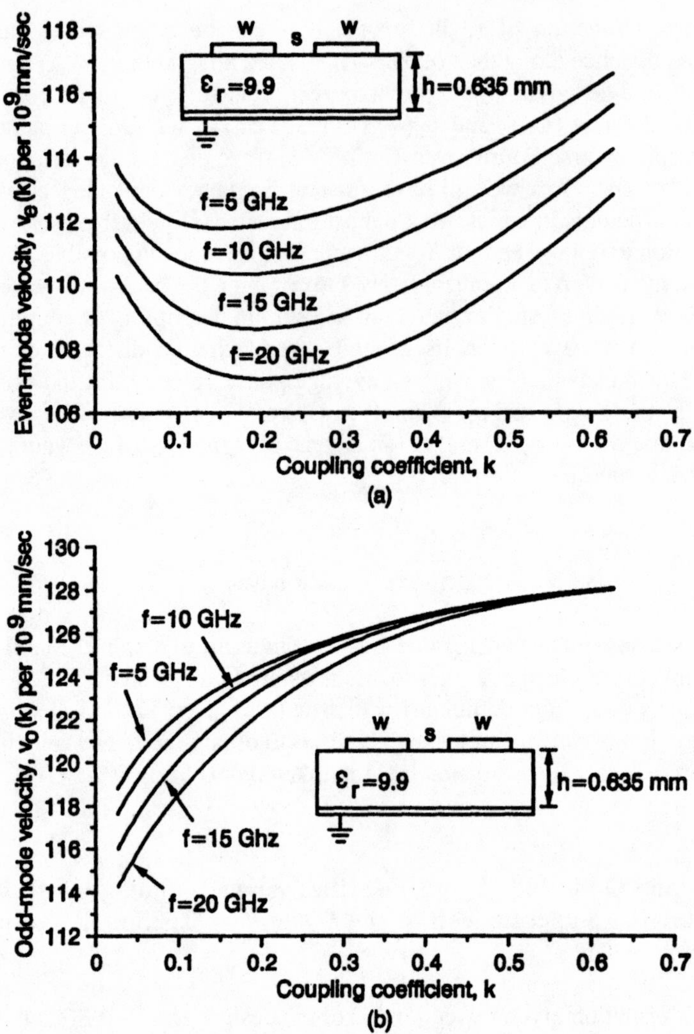

Figure 9.2 Computed phase velocities for alumina substrate with $\epsilon_r = 9.9$ and $h = 0.635$ mm: (a) even mode and (b) odd mode.

9.3.2 Even- and Odd-Mode Phase Velocities versus Coupling Coefficient for GaAs Substrate with $\epsilon_r = 12.9$ and $h = 0.2$ mm and 0.1 mm

The first set of curves are computed for $h = 0.2$ mm from $f = 5$–35 GHz in steps of 5 GHz and are shown in Figure 9.5. For $h = 0.1$ mm the results cover $f = 25$–95 GHz in steps of 10 GHz, which are given in Figure 9.6. In both cases variation in even-mode

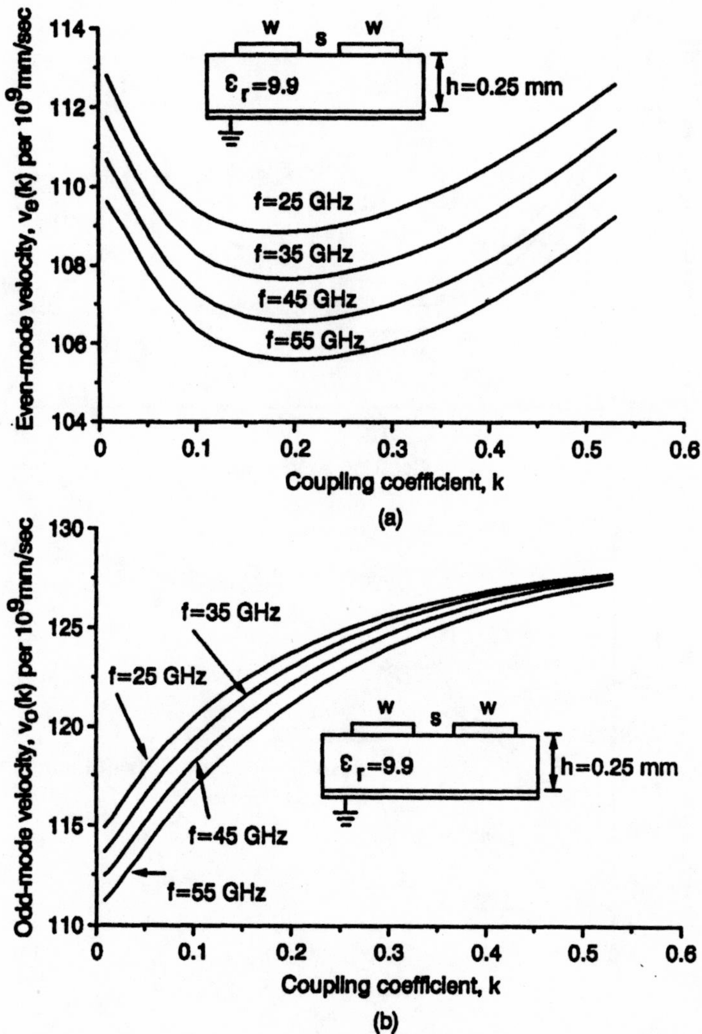

Figure 9.3 Computed phase velocities for alumina substrate with $\epsilon_r = 9.9$ and $h = 0.25$ mm: (a) even mode, (b) odd mode.

phase velocity with coupling coefficient is very small. However, variation in odd-mode phase velocity with increasing coupling coefficient is large. This variation is larger for the substrate with $h = 0.1$ mm. This is an expected result and the same is true for other substrates. We have shown in the introduction section that the two phase velocities converge to a finite value when $s \rightarrow 0$. This phase velocity limit is

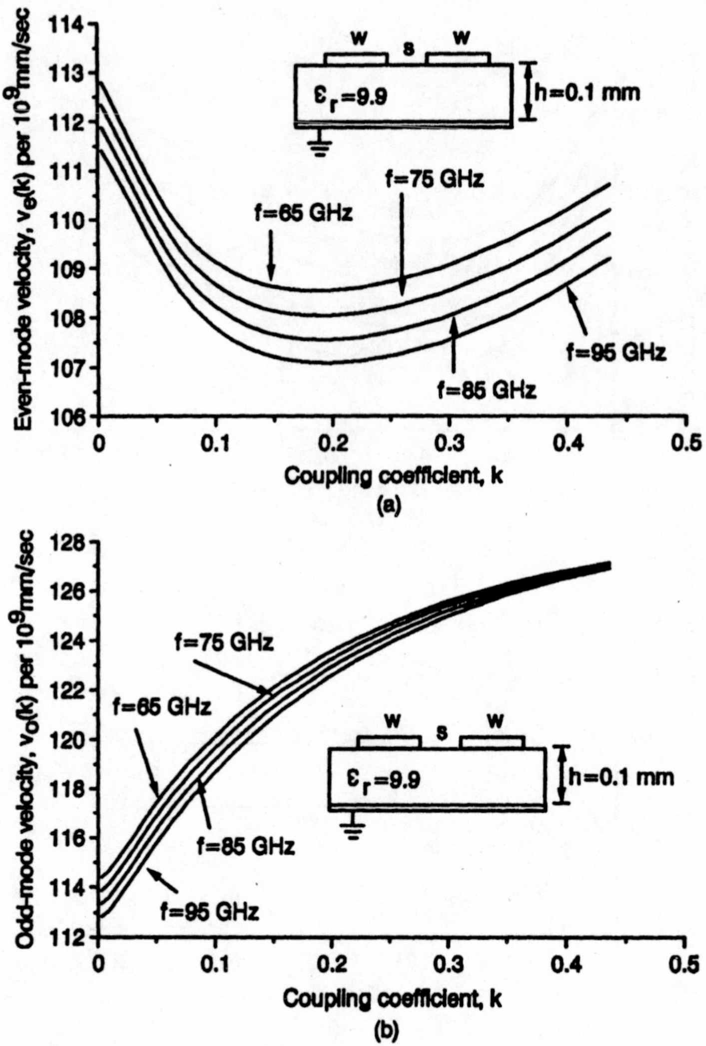

Figure 9.4 Computed phase velocities for alumina substrate with $\epsilon = 9.9$ and $h = 0.1$ mm: (a) even mode, (b) odd mode.

independent of the substrate thickness because the width of the coupled lines vanishes. Therefore, for a given substrate with a relative dielectric constant, the difference in the odd-mode phase velocity for the same amount of coupling increases with decreasing substrate thickness. This results in an increased phase velocity difference between the even and odd modes.

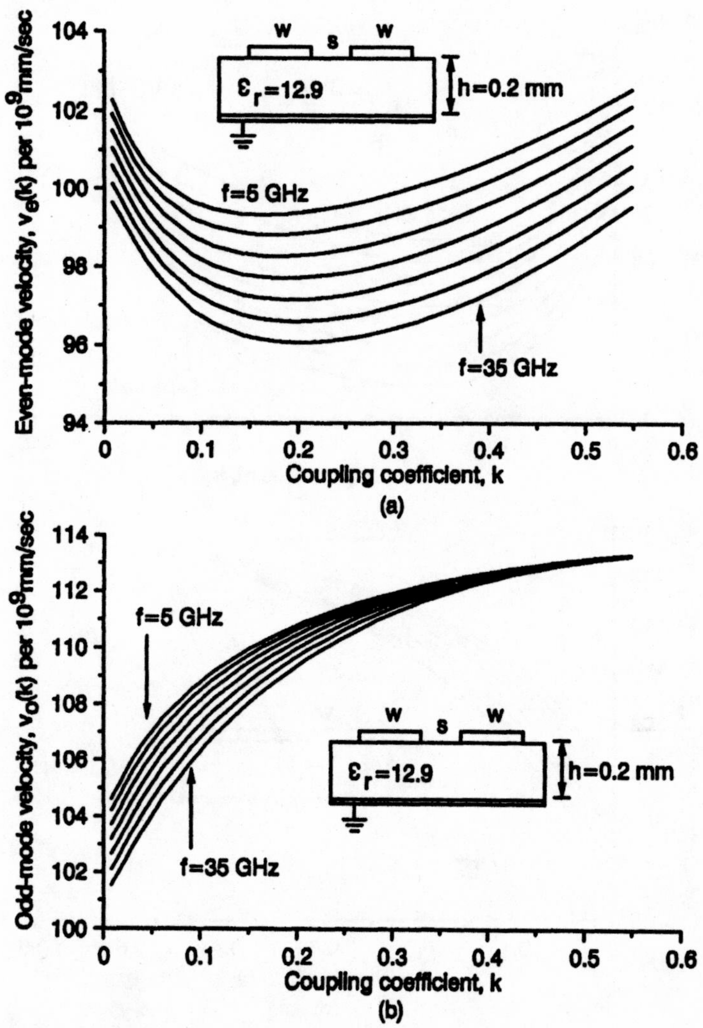

Figure 9.5 Computed phase velocities for GaAs substrate with $\epsilon_r = 12.9$ and $h = 0.2$ mm: (a) even mode, (b) odd mode.

9.3.3 Even- and Odd-Mode Phase Velocities versus Coupling Coefficient for RT-Duroid Substrate with $\epsilon_r = 10.5$ and $h = 0.2$ mm

RT-Duroid is technologically a very important substrate because its machining and handling is very easy compared to ceramic substrates. Ceramic-filled RT-Duroid has a high dielectric constant, which can be gainfully employed for circuit miniaturization.

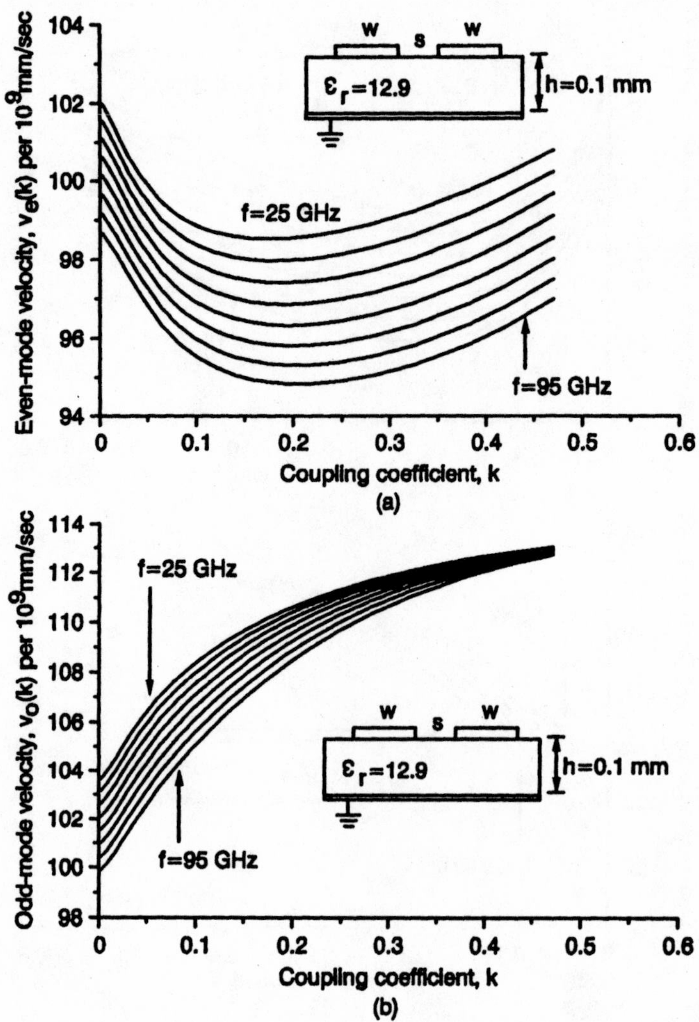

Figure 9.6 Computed phase velocities for GaAs substrate with $\epsilon_r = 12.9$ and $h = 0.1$ mm: (a) even mode, (b) odd mode.

However, the conductor layer is usually copper, which may not be suitable for circuits requiring a tight control on spacing between coupled lines. In codirectional couplers we do not have this limitation imposed on spacing, which makes RT-Duroid a preferred substrate because of its low cost. The even- and odd-mode phase velocities for a substrate thickness of 0.2 mm are computed from $f = 5$–35 GHz in steps of 5 GHz and are given in Figure 9.7.

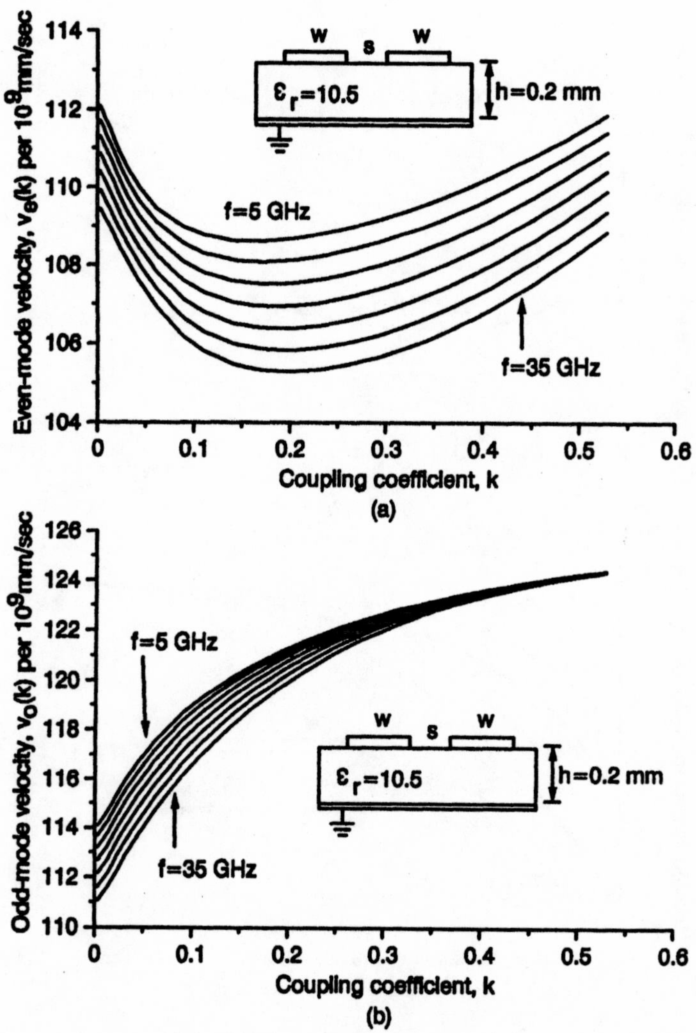

Figure 9.7 Computed phase velocities for RT-Duroid substrate with $\epsilon_r = 10.5$ and $h = 0.2$ mm: (a) even mode, (b) odd mode.

9.3.4 Even- and Odd-Mode Phase Velocities versus Coupling Coefficient for Barium Tetratitanate Substrate with $\epsilon_r = 37$ and $h = 0.635$ mm

The last set of design curves are given for $BaTi_4O_9$ with $\epsilon_r = 37$. The high dielectric constant of this substrate enables length reduction by about 40% compared to alumina substrate of the same thickness. The computed results are shown in Figure 9.8. The

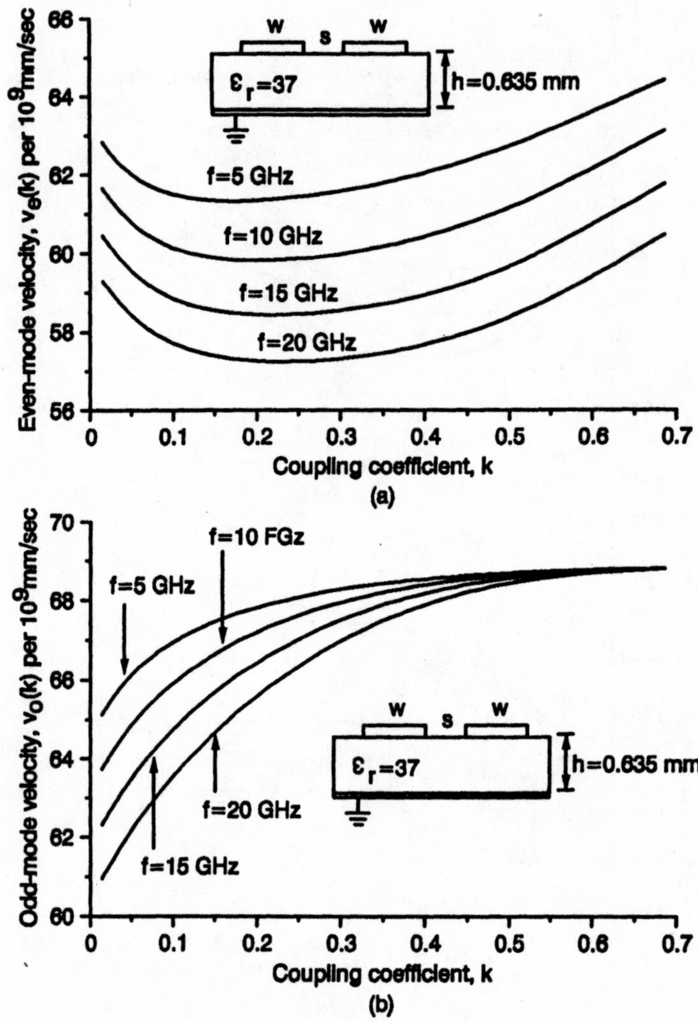

Figure 9.8 Computed phase velocities for Barium Tetratitanate substrate with $\epsilon_r = 37$ and $h = 0.635$: (a) even mode, (b) odd mode.

percentage phase velocity difference between the even and odd modes at a given k is about the same as that for alumina substrate. However, linewidths are about six times narrower due to the very high ϵ_r. At 5 GHz for $k = 0.3$, the linewidth for a 50 Ω coupler is about 0.1 mm (see Figure 3.19). It is possible to enhance the phase velocity difference by employing thinner substrates, but linewidths become very narrow, making the circuit impractical and lossy.

9.4 DESIGN EXAMPLES

Three examples are computed on 0.635-mm-thick Barium Tetratitanate substrate to illustrate the design of codirectional couplers with the theory presented so far. In all cases forward coupling will be investigated up to 20 GHz. In the first design example the coupler length is taken as $5(\lambda/4)$ at 10 GHz. The maximum coupling is limited to below 0.06. The phase velocities are at $k(-l/2)$ and $k(0)$ are obtained from Figure 9.8. We assume a linear variation in phase velocities both with ω and x.

First, a dummy backward-wave coupling channel is defined between 0 and 1 GHz. This gives the desired physical coupler to realize the computed phase velocities, which are taken as

$$v_e(x = -l/2, f = 0) = 63 \times 10^9 \text{ mm/s} \qquad v_o(x = -l/2, f = 0) = 65 \times 10^9 \text{ mm/s}$$
$$v_e(x = 0, f = 0) = 62 \times 10^9 \text{ mm/s} \qquad v_o(x = 0, f = 0) = 66 \times 10^9 \text{ mm/s}$$
$$v_e(x = -l/2, f = 20) = 59 \times 10^9 \text{ mm/s} \qquad v_o(x = -l/2, f = 20) = 61 \times 10^9 \text{ mm/s}$$
$$v_e(x = 0, f = 20) = 58 \times 10^9 \text{ mm/s} \qquad v_o(x = 0, f = 20) = 62 \times 10^9 \text{ mm/s}$$

We then define the functions $v_e(\omega, x)$ and $v_o(\omega, x)$ as follows:

$$v_e(\omega, x) = (-2x/l - f/5 + 63) \times 10^9 \text{ mm/s}$$
$$v_o(\omega, x) = (-2x/l - f/5 + 65) \times 10^9 \text{ mm/s}$$

These are analytically simple functions that are then used to obtain

$$\int_{-l/2}^{l/2} \frac{dx}{v_e(\omega, x)} = l \ln\left[\frac{5 - f + 315}{-f + 315}\right]$$
$$\int_{-l/2}^{l/2} \frac{dx}{v_o(\omega, x)} = l \ln\left[\frac{5 - f + 325}{-f + 325}\right]$$

We then use equations (9.13) to (9.16) to compute the S-parameters with l/v_e and l/v_o replaced by these integrals. The computed results are shown in Figure 9.9 A -3 dB forward-wave coupling is obtained at 19.6 GHz.

In the second design, coupler length is kept the same as in the first design but a higher dummy coupling is used to increase $k(0)$. The corresponding phase velocities are again obtained from Figure 9.8:

$$v_e(x = 0, f = 0) = 62 \times 10^9 \text{ mm/s} \qquad v_o(x = 0, f = 0) = 68 \times 10^9 \text{ mm/s}$$
$$v_e(x = 0, f = 20) = 57.5 \times 10^9 \text{ mm/s} \qquad v_o(x = 0, f = 20) = 66 \times 10^9 \text{ mm/s}$$

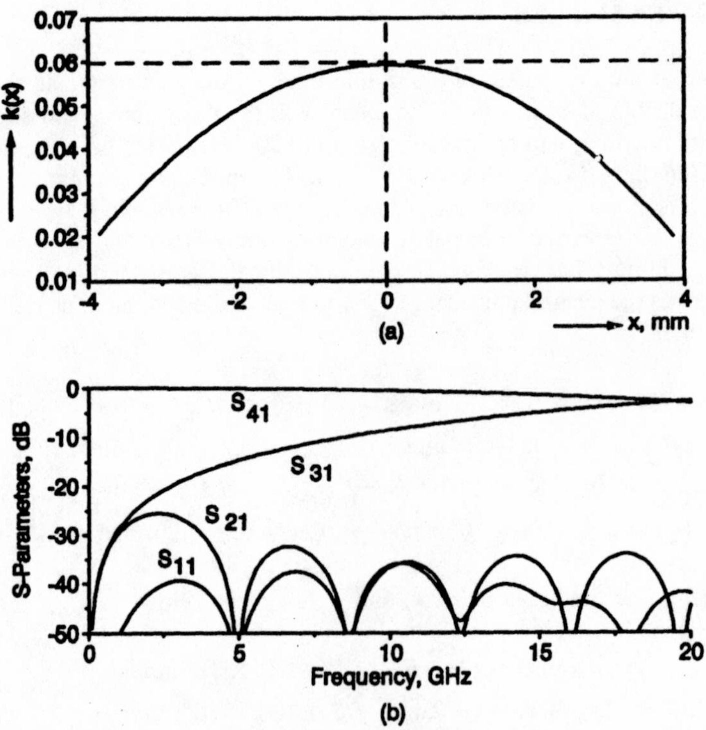

Figure 9.9 Computed results for a nonuniform codirectional coupler with $l = 5(\lambda/4)$ at 10 GHz and very large coupled line spacing on $BaTi_4O_9$: (a) continuous-coupling coefficient and (b) performance in the 0–20 GHz range.

Again, we assume linear functions for $v_e(\omega, x)$ and $v_o(\omega, x)$. The computed results for this example are given in Figure 9.10. With the same coupler length, forward-wave coupling is increased due to the increase in phase velocity difference. The maximum coupling coefficient is 0.257. A -3 dB coupling is obtained at 12.5 GHz with a $5(\lambda/4)$-long coupler at 10 GHz.

We see that isolation is below -30 dB in a wide frequency range of the forward-wave coupling. If we had used a uniform coupler with $k = 0.257$ and the same length, the performance would have deteriorated sharply, because backward-wave coupling in uniform couplers repeats itself at $3f_c$. And because we have backward-wave coupling, the difference between even- and odd-mode phase velocities causes a high input reflection. In the case of uniform couplers, G_e and G_o can be replaced by the following:

$$G_e = \frac{1}{2}\ln\left\{\frac{1+k}{1-k}\right\}\sin(\theta_e) \tag{9.21}$$

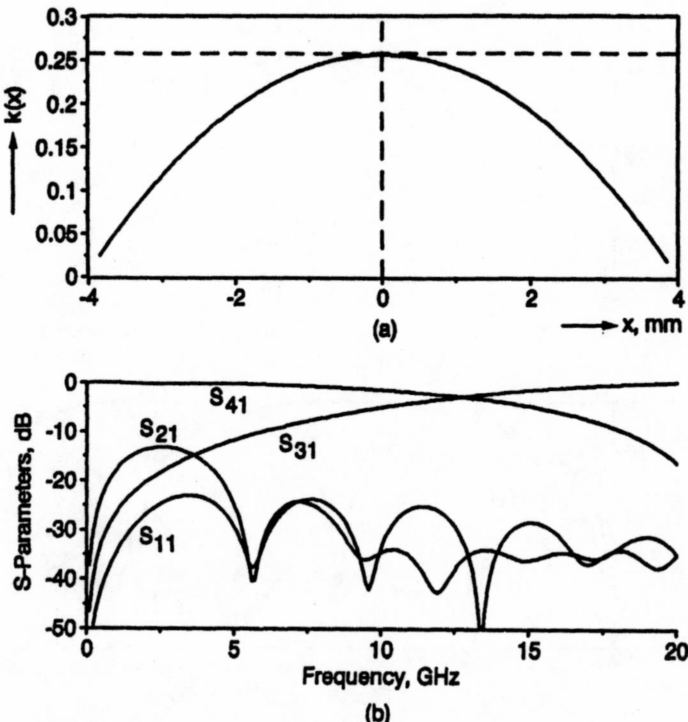

Figure 9.10 Computed results for a nonuniform codirectional coupler with $l = 5(\lambda/4)$ at 10 GHz and moderate coupled line spacing on $BaTi_4O_9$: (a) continuous-coupling coefficient and (b) performance in the 0–20 GHz range.

$$G_o = \frac{1}{2}\ln\left\{\frac{1+k}{1-k}\right\}\sin(\theta_o) \tag{9.22}$$

where θ_e and θ_o are the even- and odd-mode electrical lengths.

In the last design example given in this section, we increase the coupler length from $5l_c$ to $15l_c$ but keep the same phase velocities of the previous example. The computed results are shown in Figure 9.11. As expected, forward coupling increases with increasing length. We now have -3 dB coupling at $f = 4.5$ GHz.

For given phase velocities we can actually deduce the approximate coupler length required for a specified S_{31} by using the simplified eq. (9.17). For instance, if we require $S_{31} = 0.707$ at 10 GHz, we then have

$$l = \frac{v_e v_o}{(v_e - v_o)\omega}\sin^{-1}(S_{31}) = \frac{v_e v_o}{80(v_e - v_o)}$$

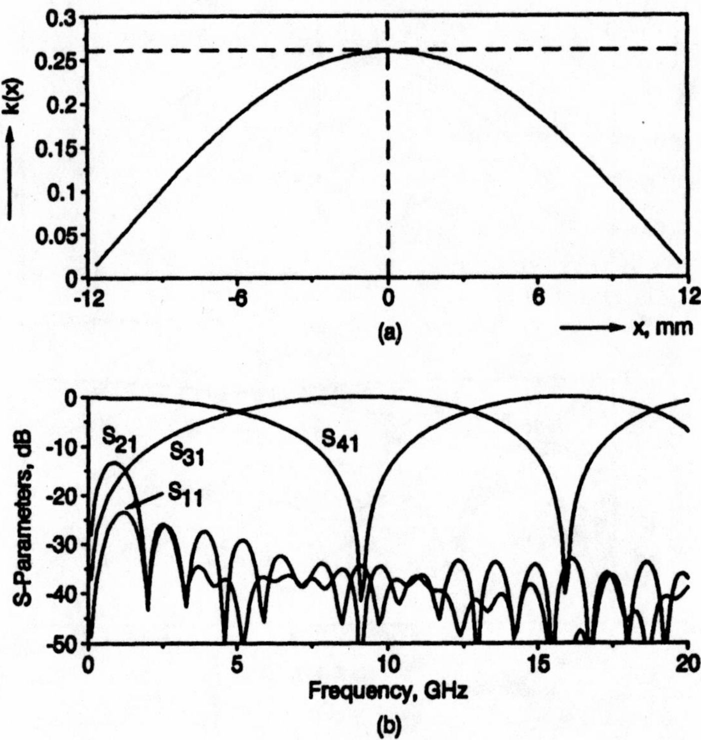

Figure 9.11 Computed results for a nonuniform codirectional coupler with $l = 15(\lambda/4)$ at 10 GHz and moderate coupled line spacing on $BaTi_4O_9$: (a) continuous-coupling coefficient and (b) performance in the 0–20 GHz range.

This is only an approximate result because we have neglected the variation in phase velocities with respect to x. However, it is still useful to have an indication (rough estimate) of the required length.

9.5 WIGGLY CODIRECTIONAL COUPLERS

Wiggling in nonuniform codirectional couplers can be introduced in two ways: (1) wiggling the outer edges and (2) wiggling the inner edges. For loose coupling values, the former reduces both even- and odd-mode phase velocities by about the same amount. Sugiura [1] has reported results for uniform couplers with slots (rectangular) at outer edges and inner edges. His results show that slotting the outer edges reduces even-mode velocity significantly for a given spacing. Odd-mode velocity is also reduced, but the effect of slotting the outer edges is less for reduced spacing between the coupled lines. We have predicted and experimentally verified that for phase-velocity compensated nonuniform directional couplers, wiggling the inner edges with loose couplers reduces both the even and odd modes. For loose coupling, wiggling the outer edges introduces the same effect with the exception that the phase velocity difference remains almost constant. Because the bandpass-coupling principle allows us to design nonuniform double-coupled codirectional couplers with any practically realizable gap, we can take advantage of the increased phase velocity difference with decreasing spacing thereby realizing codirectional couplers with reduced lengths for any desired coupling at a given frequency. This length reduction is obtained as compared to codirectional couplers with no wiggle.

Because they can easily achieve 0 dB coupling, codirectional couplers can be used as passive 90° phase shifters. As it was stated earlier the sign of the phase quadrature in these couplers depends on the dominant phase velocity. If the even-mode phase velocity is larger we have angle (coupled) − angle (direct) = −90°. Wiggling the inner edges of the coupled lines reduces the odd-mode velocity more than the even-mode velocity. At a specific value of wiggle depth, for a given spacing, the two mode velocities become equal. Increasing the wiggle depth further reduces the phase velocities but the effect on the odd mode is much larger for reduced spacing. In this case the phase quadrature becomes angle (coupled) − angle (direct) = 90°. These phase quadrature characteristics of codirectional couplers provide significant design flexibility with some applications, which will be discussed in the next chapter.

9.5.1 Design Examples

Semiempirical design curves for nonuniform codirectional couplers with wiggly inner edges have been reported for alumina substrate with $\epsilon_r = 9.9$ and $h = 0.635$ mm [2]. The first design example for wiggly codirectional couplers is given for alumina substrate with $\epsilon_r = 9.9$ and $h = 0.635$ mm. Forward coupling is investigated in the 0–20 GHz band with a coupler length of 18.8. The phase velocities for this design are assumed to be linear functions from $-l/2$ to 0. At these points the following velocities are used:

$$v_e(x = -l/2, f = 0) = 108 \times 10^9 \text{ mm/s} \qquad v_e(x = -l/2, f = 20) = 106 \times 10^9 \text{ mm/s}$$
$$v_o(x = -l/2, f = 0) = 105 \times 10^9 \text{ mm/s} \qquad v_o(x = -l/2, f = 20) = 104 \times 10^9 \text{ mm/s}$$
$$v_e(x = 0, f = 0) = 112 \times 10^9 \text{ mm/s} \qquad v_e(x = 0, f = 20) = 107 \times 10^9 \text{ mm/s}$$
$$v_o(x = 0, f = 0) = 102 \times 10^9 \text{ mm/s} \qquad v_o(x = 0, f = 20) = 101 \times 10^9 \text{ mm/s}$$

The computed results are given in Figure 9.12. Maximum coupling coefficient is 0.28 at the coupler center, which gives the specified phase velocities at $x = 0$. The

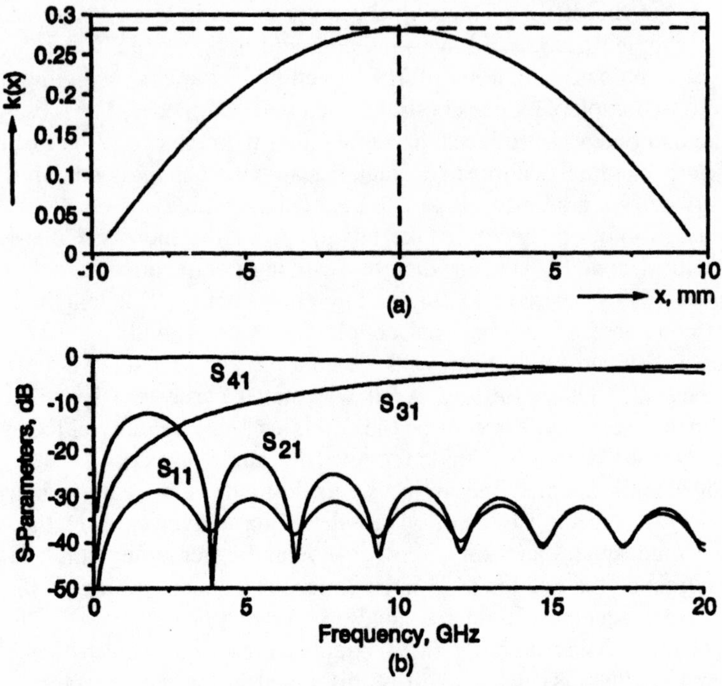

Figure 9.12 Computed results for a 18.8-mm-long nonuniform wiggly coupler on 0.635-mm-thick alumina substrate: (a) continuous-coupling coefficient and (b) performance in the 0–20 GHz range.

computed performance shows −3 dB forward-wave coupling at 16 GHz with isolation better than −30 dB.

This design is built on a 25.4 × 25.4 × 0.635 mm^3 alumina substrate. The realized circuit is shown in Figure 9.13. Its measured results are given in Figure 9.14. We see a shift in the frequency of the −3 dB coupling point. This is due to the difficulty in predicting the exact phase velocities with directional couplers. Another factor is the selected phase velocity functions. We have assumed that these functions are linear. However, for smaller coupler lengths they become increasingly nonlinear. Small errors in calculating the wiggly depths to achieve linear phase velocities reflect in the measured results as a shift in the frequency for a desired coupling at a given frequency. Isolation is slightly higher than the predicted value due to small mismatches at the transition points from wiggle coupled lines to the connecting 50-Ω lines. At the vicinity of the −3 dB coupling region, isolation is less than −30 dB. Input reflection is maintained below −15 dB in the 0–20 GHz range. The phase quadrature given for angle (S_{41}) − angle (S_{31}) is good with less than ±1.5° deviation from 15 to 18 GHz.

Figure 9.13 The nonuniform wiggly codirectional coupler on a 25.4 × 25.4 × 0.635 mm^3 alumina substrate.

A second design with a 29.3-mm coupler length is again designed on alumina substrate with the same phase velocities used in the first design. The computed results for this coupler are given in Figure 9.15. A 0 dB coupling is indicated at 6.6 GHz. Because the phase velocities are assumed to be linear, the second 0 dB coupling point occurs at 6.6 × 3 = 19.8 GHz. The measured results are shown in Figure 9.16. In this case, a better agreement is observed between the computed and measured results, because the assumption in taking the phase velocities as linearly varying functions holds better for longer couplers as the taper rate is smaller. Input reflection and isolation are around −20 dB and −30 dB, respectively. The sign reversal in the measured phase quadrature is an anomaly of the network analyzer.

9.6 WIDEBAND CODIRECTIONAL COUPLERS

In codirectional couplers, forward-wave coupling increases until it reaches 0 dB. If the phase velocity difference is constant, forward-wave coupling disappears at twice the frequency of the first 0 dB coupling. This characteristic of codirectional couplers limits the usable bandwidth to less than 25%. Wideband codirectional coupling is possible if we have different phase constants for the two coupled lines [3]. This requires the use of an asymmetrical configuration with two different widths for the coupled lines. For practical reasons, we also must change the impedance level of the design. Matching

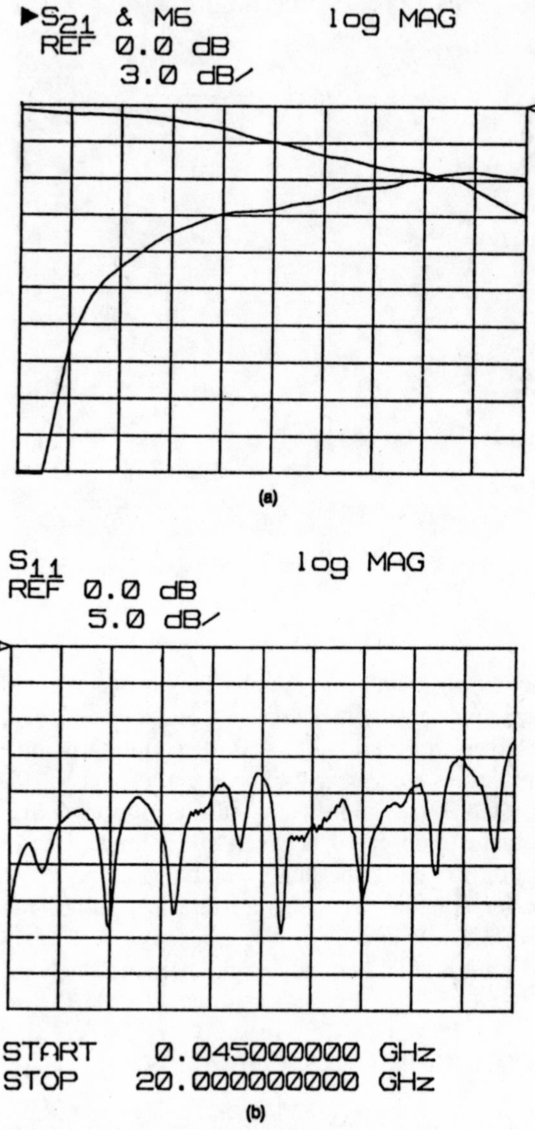

Figure 9.14 Measured results for the nonuniform wiggly codirectional coupler with l = 18.8 mm: (a) performance in the 0–20 GHz range. (b) input reflection.

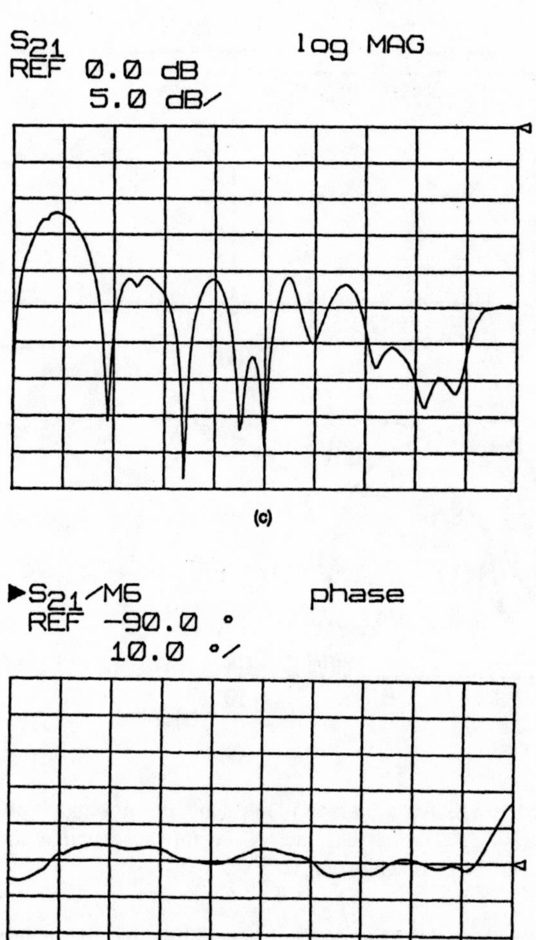

START 0.045000000 GHz
STOP 20.000000000 GHz

(d)

Figure 9.14 (continued) Measured results for the nonuniform wiggly codirectional coupler with l = 18.8 mm: (c) isolation, (d) phase quadrature (angle(S_{31}) − angle(S_{41})).

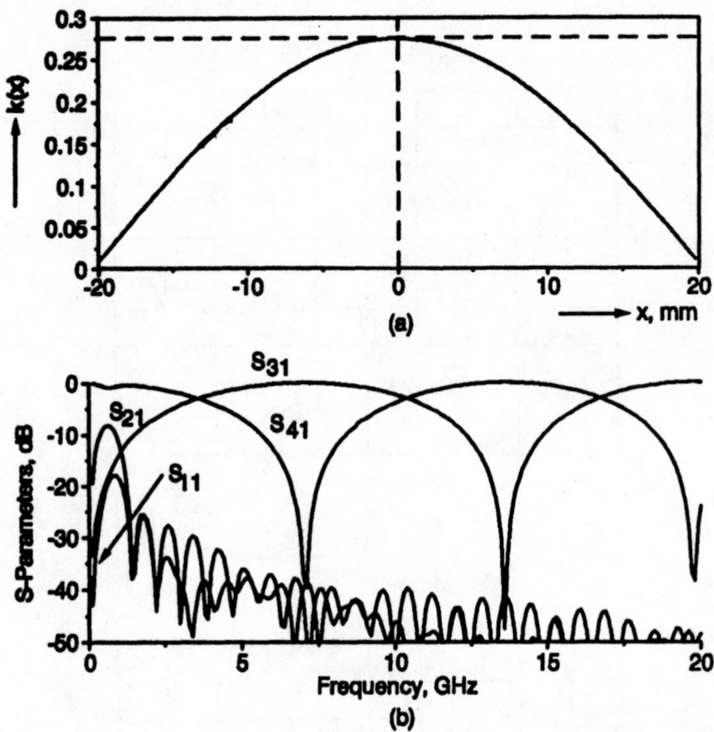

Figure 9.15 Computed results for a 39.2-mm-long nonuniform wiggly codirectional coupler on 0.635-mm-thick alumina substrate: (a) continuous-coupling coefficient and (b) performance in the 0–20 GHz range.

sections are then used to obtain the desired impedance for the system of asymmetrical coupled lines.

Codirectional couplers with wideband performance have also been reported by other authors [4–6]. Both Gunton [4,5] and Islam [6] make use of the forward-wave coupling principle based on coupled comblines. The coupled comblines have different phase constants due to different finger lengths.

However, we know that the phase velocities in nonuniform couplers are functions of frequency and the coupling coefficient. Therefore, the coupled signal can be expressed as

$$S_{31} = \sin\left[\omega\left(\int_0^l \frac{1}{v_e(\omega, x)} - \frac{1}{v_o(\omega, x)} \mathrm{d}x\right)\right] \tag{9.23}$$

Wideband performance can be achieved if the even- and odd-mode phase velocities are similar to those given in Figure 9.17. This figure shows that we must have equal phase velocities at f_2 while we maintain the phase velocity difference at a lower frequency. Therefore, the argument of the sine function of the coupled signal should give a similar function as shown in Figure 9.17.

A close inspection of Figure 9.17 suggests the use of a phase velocity compensation technique. We can use the dispersive nature of microstrip to realize the phase velocity functions given in Figure 9.17. In backward-wave couplers we introduce phase velocity compensation at the design center frequency, thereby limiting isolation in a wide bandwidth. In codirectional couplers, we introduce phase velocity compensation at the upper end of the frequency range. The amount of dispersion for a given substrate and frequency range can be deduced from the design curves given in Figures 9.2 through 9.8.

9.6.1 Design Example

A wideband codirectional coupler is designed on a 0.635-mm-thick alumina substrate. Forward-wave coupling is investigated in the 0–20 GHz range. The phase velocities are assumed to be of the type given in Figure 9.17 with the following end values:

$$v_e(x = -l/2, f = 0) = 105 \times 10^9 \text{ mm/s} \quad v_e(x = -l/2, f = 20) = 103 \times 10^9 \text{ mm/s}$$

$$v_o(x = -l/2, f = 0) = 103 \times 10^9 \text{ mm/s} \quad v_o(x = -l/2, f = 20) = 101 \times 10^9 \text{ mm/s}$$

$$v_e(x = 0, f = 0) = 112.5 \times 10^9 \text{ mm/s} \quad v_e(x = 0, f = 20) = 107 \times 10^9 \text{ mm/s}$$

$$v_o(x = 0, f = 0) = 107 \times 10^9 \text{ mm/s} \quad v_o(x = 0, f = 20) = 106.9 \times 10^9 \text{ mm/s}$$

The coupler length is 5 wavelengths long at 10 GHz. The computed results are shown in Figure 9.18. The continuous-coupling coefficient has a maximum value of 0.27, which gives the desired phase velocities at $x = 0$. The coupling balance is ± 1 dB in the 10–18 GHz range. The coupling balance is $\pm$ 0.4 dB in the 11.5–17 GHz range.

As the results suggest, it is sufficient to introduce phase velocity equalization at a high frequency. However, the phase velocities in the, say, 10–18 GHz range are not equal. They have a decreasing difference toward 18 GHz. This implies that ultrawideband performance is possible by increasing the coupler length and adjusting the wiggle depth. By maintaining the desired relationship between the phase velocities a high-pass-type coupling is achievable.

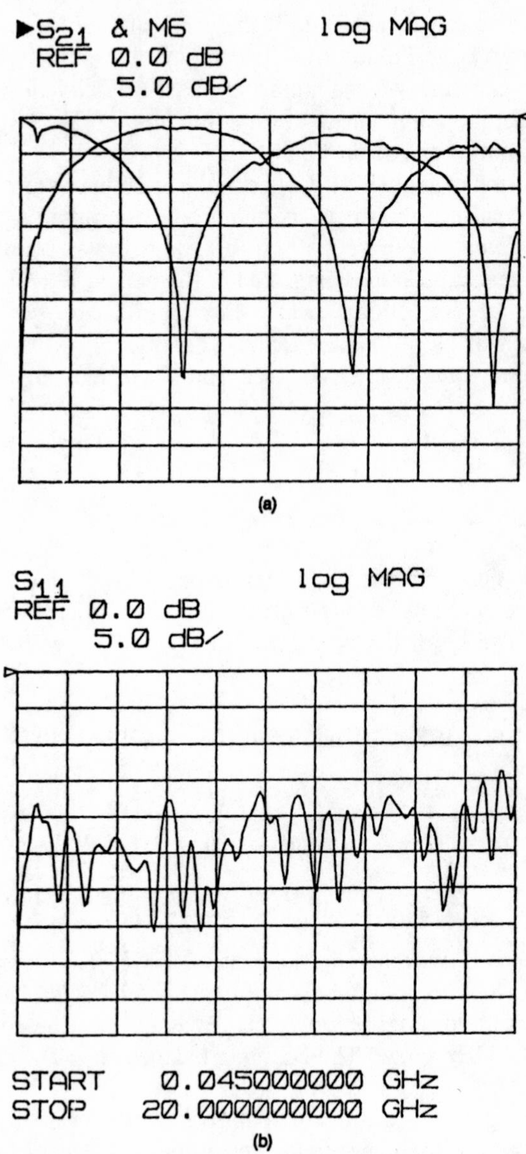

Figure 9.16 Measured results for the nonuniform wiggly codirectional coupler with l = 39.2 mm: (a) performance in the 0–20 GHz range, (b) input reflection.

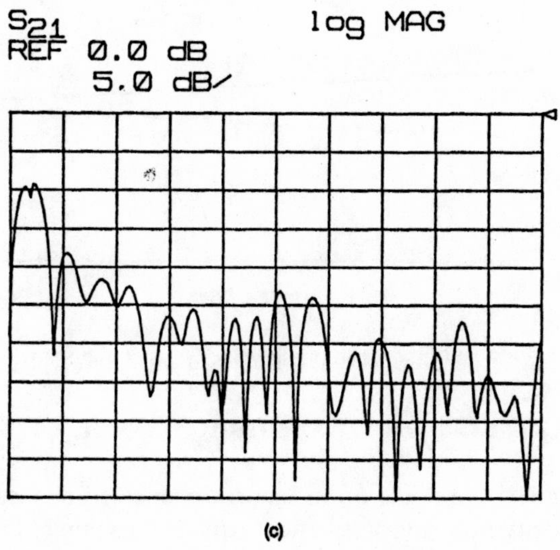

(c)

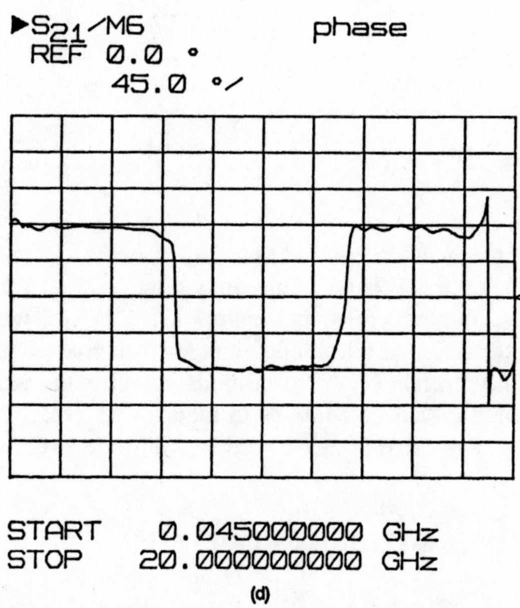

(d)

Figure 9.16 (continued) Measured results for the nonuniform wiggly codirectional coupler with l = 39.2 mm: (c) isolation, (d) phase quadrature (angle(S_{41}) − angle(S_{31})).

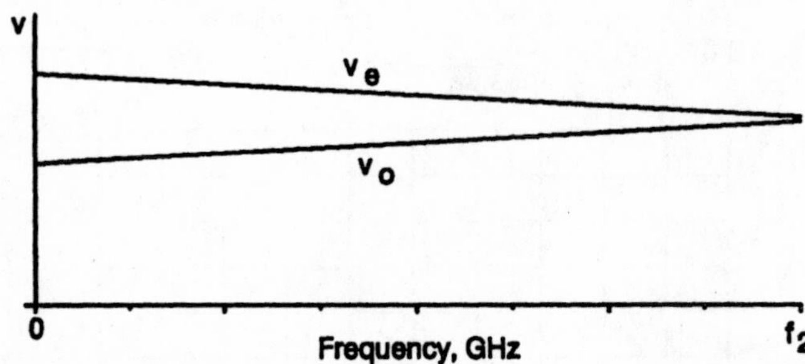

Figure 9.17 Required even- and odd-mode phase velocity functions for wideband performance.

9.7 TANDEM CONNECTION OF CODIRECTIONAL COUPLERS

Forward-wave couplers are made up of tandem-connected elemental sections. With complex phase velocity functions it is better to divide the coupler into several sections. The coupled signal is then given by

$$S_{31} = \sin\left[\omega\left(\int_0^{l_1} \frac{1}{v_{e_1}(\omega, x)} - \frac{1}{v_{o_1}(\omega, x)} dx\right) + \omega\left(\int_{l_1}^{l_2} \frac{1}{v_{e_2}(\omega, x)} - \frac{1}{v_{o_2}(\omega, x)} dx\right)\right]$$
$$+ \omega\left(\int_{l_2}^{l_3} \frac{1}{v_{e_3}(\omega, x)} - \frac{1}{v_{o_3}(\omega, x)} dx\right) + \cdots + \omega\left(\int_{l_{n-1}}^{l_n} \frac{1}{v_{e_n}(\omega, x)} - \frac{1}{v_{o_n}(\omega, x)} dx\right) \tag{9.24}$$

A typical example in which the odd-mode phase velocity experiences a sharp change is given in Figure 9.19. This coupler may have $l_1 = l_2$. An illustrative phase velocity plot along the coupler is given in Figure 9.19(b). For the wiggly section even-mode velocity is faster than the odd-mode velocity; because wiggling for relatively small gaps has very little effect on the even-mode velocity, the wiggle depth toward the end of the first section can be adjusted to reduce or increase the phase velocity difference. For the smooth-edge coupled section, odd-mode velocity is much faster than the even-mode velocity.

9.8 CASCADED CODIRECTIONAL COUPLERS

Two couplers may be cascaded by connecting the coupled port of the second coupler to the isolated port of the first coupler. It is, therefore, not possible to cascade codirectional couplers directly: delay lines and crossovers are required to make the necessary connections. A possible way of cascading codirectional couplers is given in Figure 9.20. With an even number of cascaded couplers, the coupled and direct ports are on opposite sides. The output ports are adjacent with an odd number of couplers.

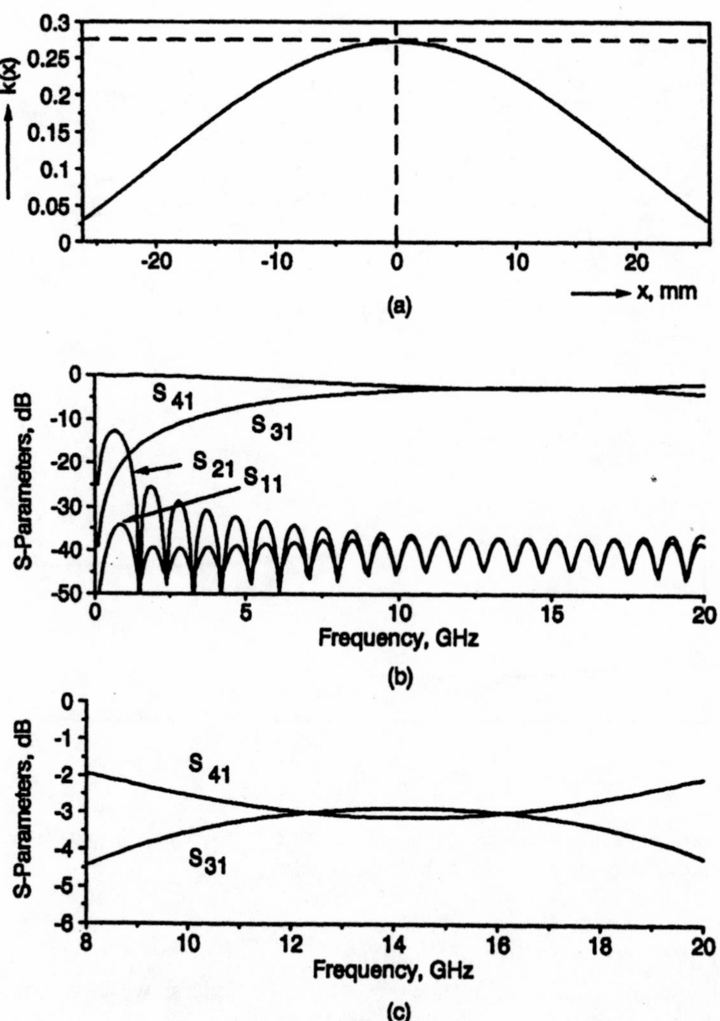

Figure 9.18 Computed results for a wideband nonuniform wiggly codirectional coupler on alumina substrate: (a) continuous-coupling coefficient, (b) performance in the 0–20 GHz range, and (c) performance in the design bandwidth.

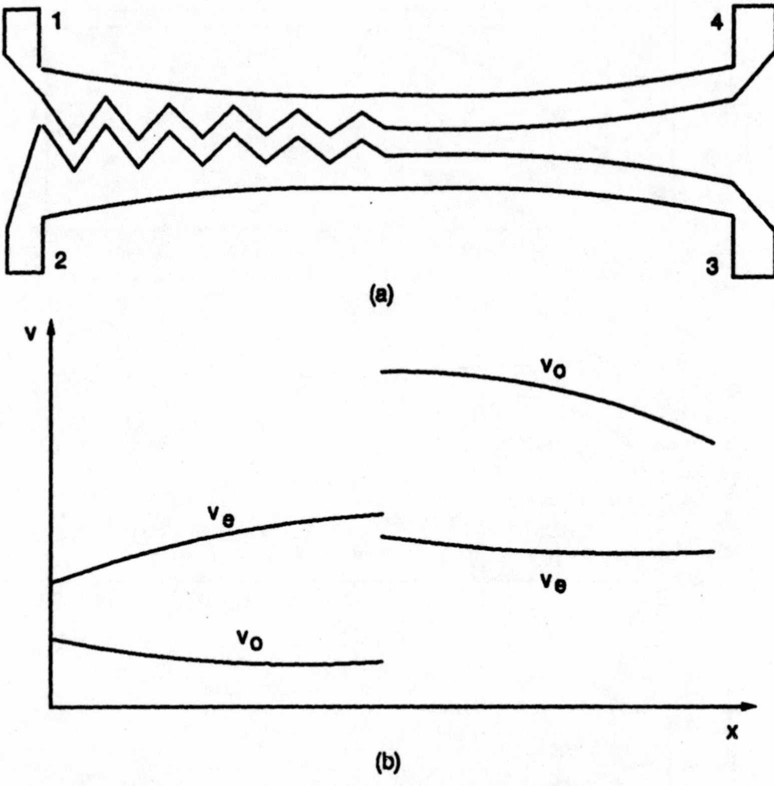

Figure 9.19 (a) Codirectional coupler configuration for a sharp change in phase velocities. (b) Illustrative phase velocities along the coupler.

For the analysis of cascaded codirectional couplers, we first consider two sections. The coupled signal is obtained by

$$S_{31} = \sin\left[\omega\left(\int_0^{l_1} \frac{1}{v_{e_1}(\omega, x)} - \frac{1}{v_{o_1}(\omega, x)} \mathrm{d}x\right)\right] + \cos^2\left[\omega\left(\int_0^{l_1} \frac{1}{v_{e_1}(\omega, x)} - \frac{1}{v_{o_1}(\omega, x)} \mathrm{d}x\right)\right] \times \sin\left[\omega\left(\int_0^{l_2} \frac{1}{v_{e_2}(\omega, x)} - \frac{1}{v_{o_2}(\omega, x)} \mathrm{d}x\right)\right] \mathrm{e}^{j2\beta d} \tag{9.25}$$

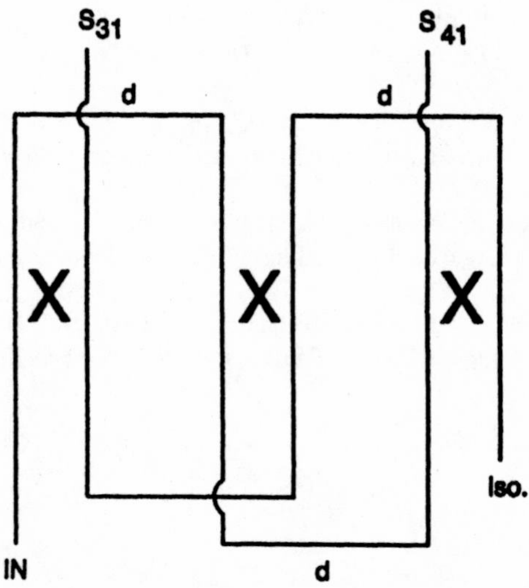

Figure 9.20 Cascaded codirectional couplers.

For three sections we have

$$S_{31} = [S_{31}]_{2-\text{section}} + \cos^2\left[\omega\left(\int_0^{l_1} \frac{1}{v_{e_1}(\omega, x)} - \frac{1}{v_{o_1}(\omega, x)}\,dx\right)\right] + \cos^2$$
$$\times\left[\omega\left(\int_0^{l_2} \frac{1}{v_{e_2}(\omega, x)} - \frac{1}{v_{o_2}(\omega, x)}\,dx\right)\right] \sin\left[\omega\left(\int_0^{l_3} \frac{1}{v_{e_3}(\omega, x)} - \frac{1}{v_{o_3}(\omega, x)}\,dx\right)\right] e^{j4\beta d} \tag{9.26}$$

The performance for any number of cascaded sections can be derived by observing the periodicity in the preceding equations. For every new added section there is a cosine function and the performance of n section can be expressed in terms of the performance derived for the previous $n - 1$ sections.

REFERENCES

[1] Sugiura, T., "Analysis of Distributed-Lumped Strip Lines," *IEEE Trans. on Microwave Theory and Tech.*, Vol. MTT-25, 1977, pp. 656–661.

[2] Uysal, S., J. Watkins, and C. W. Turner, "Sum-Difference Circuits Using 0 dB and −3 dB Co-Directional Couplers for Hybrid Microwave and MIMIC Circuit Applications," *IEEE MMT-S Int. Microwave Symp. Digest,* 1992, pp. 937–940.

[3] Ikalainen, P. K., and G. L. Matthaei, ''Wideband Forward-Coupling Microstrip Hybrids with High Directivity,'' *IEEE Trans. on Microwave Theory and Tech.,* Vol. MTT-35, 1987, pp. 719–725.

[4] Gunton, D. J., and E. G. S. Paige, ''Directional Coupler for Gigahertz Frequencies, Based on the Coupling Properties of Two Planar Comb Transmission Lines,'' *Electron. Lett.,* Vol. 11, 1975, pp. 406–408.

[5] Gunton, D. J., ''Design of Wideband Codirectional Couplers and Their Realization at Microwave Frequencies Using Coupled Comblines,'' *IEE J. Microwave Opt. and Acoust.,* Vol. 2, 1978, pp. 19–30.

[6] Islam, S., ''A New Analytic Design Technique for Two- and Three-Way Warped Mode Combline Directional Couplers,'' *IEEE Trans. on Microwave Theory and Tech.,* Vol. MTT-39, 1989, pp. 34–42.

Chapter 10
Applications of Nonuniform Line Directional Couplers

10.1 INTRODUCTION

Numerous components and subsystems use directional couplers to perform specific functions. These functions are usually derived from three main characteristics of directional couplers among their four ports:

1. Amplitude,
2. Phase, and
3. Isolation.

Some of the applications may use just one or two combinations of these characteristics. However, the most important applications utilize all three relationships. In such applications the overall component or subsystem performance depends critically on the amplitude and phase balance and isolation of the directional coupler.

In the previous chapters we described the operations and capabilities of three major directional coupling structures based on the synthesis of continuous coupling coefficient of nonuniform coupled lines. These are

1. Ultrawideband backward-wave directional couplers,
2. Bandpass-type backward-wave directional couplers, and
3. Forward-wave (codirectional) couplers.

In this chapter, we attempt to highlight some of the major applications utilizing these couplers at microwave and millimeter wave frequencies. For the benefit of the readers, any distinctive advantage arising from the use of any particular type of coupler will be given in the design.

10.2 FIXED ATTENUATOR, CROSSOVER, AND dc-BLOCK

One of the simplest use of a directional coupler is as an attentuator. In this application, the isolated port and either the direct or the coupled port are terminated. This simple circuit, where the output is obtained from the coupled port is shown in Figure 10.1. Ultrawideband high attenuation can be obtained by using double-coupled nonuniform lines. The attenuation level is given directly by the coupled signal level. That is, for 20 dB attenuation a −20 dB coupler is used. The losses, especially toward the upper end of the design bandwidth can be compensated for by specifying a linear coupling function instead of a constant one.

A crossover of two signals can easily be achieved by the use of a 0 dB directional coupler, as shown in Figure 10.2. The signals *A* and *B* entering the circuit emerge from diagonally opposite ports. Thus, the 0 dB coupler allows two lines to cross each other without dc connection. The isolation between the lines depends on the type of coupler used. For narrow to moderate bandwidth applications, codirectional couplers provide in excess of 30 dB isolation with minimum signal loss. All the types of directional couplers presented in this book provide dc isolation between an input signal and the coupled signal.

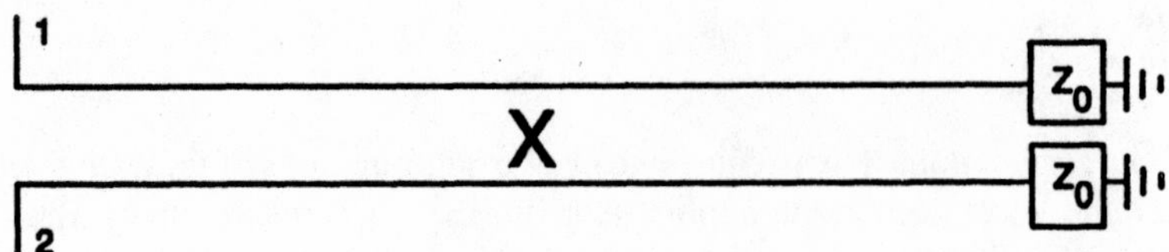

Figure 10.1 Directional coupler used as a fixed attenuator.

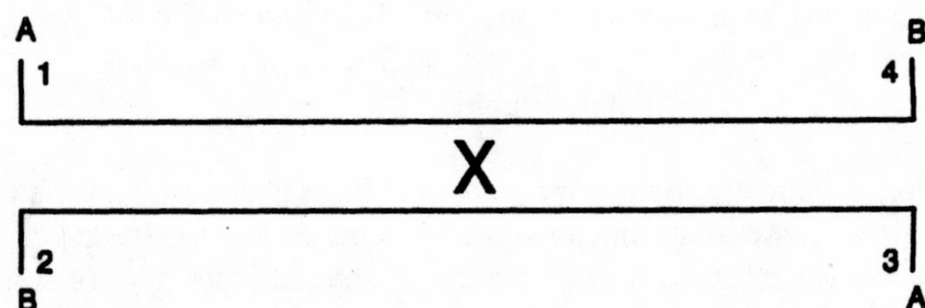

Figure 10.2. A 0 dB coupler used as a crossover.

10.3 FIXED PHASE SHIFTER

Any amount of differential phase shift can be achieved by directional couplers. The phase difference between the coupled and direct signals of a codirectional coupler with identical coupled lines is always 90° irrespective of any longitudinal asymmetry. Therefore, a 0 dB codirectional coupler can be used as a 90° phase shifter circuit, as shown in Figure 10.3.

In nonuniform backward-wave directional couplers the phase difference between the coupled and direct signals depends on the configuration. A symmetrical configuration (assume identical lines) gives a 90° phase difference between these ports if isolation and input match are perfect. This phase quadrature worsens with decreasing isolation. Phase velocity compensation ensures minimum deviation (usually less than ±5° in ultrawideband design) from phase quadrature. Therefore, 0 dB backward-wave couplers can again provide 90° fixed phase shift with respect to a reference line.

Any longitudinal asymmetry introduced into the backward-wave couplers results in a deviation from phase quadrature, because the reflection coefficient distribution function will no longer be an odd function. This also means that the Fourier transform pair given for $p(x)$ and $C(\omega)$ is no longer valid. The asymmetrical configuration is shown in Figure 10.4.

For the asymmetrical coupler, the reflection coefficient distribution function is

$$p(x) = -\frac{2}{\pi v}\int_{0}^{2\omega_c} e^{-j2\beta l} \tanh^{-1}[C(\omega)]\, d\omega \tag{10.1}$$

The new phase difference between the coupled and direct signals depends on the amount of length reduction compared to a symmetrical configuration.

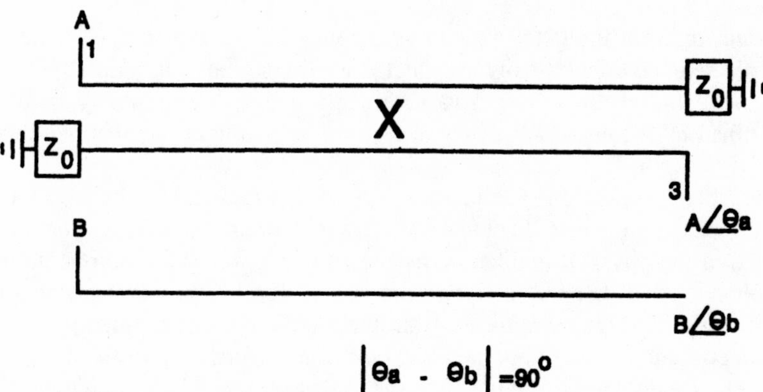

Figure 10.3 A 90° phase-shifter configuration.

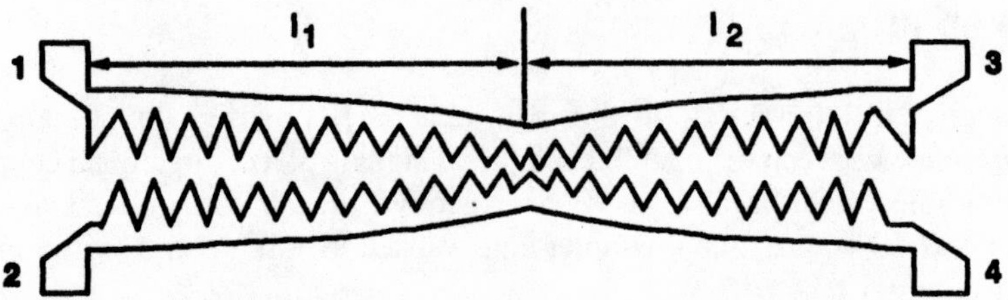

Figure 10.4 Asymmetrical backward-wave coupler used as a differential phase shifter.

An alternative fixed phase-shifter configuration can be obtained by joining the direct port and isolated port of a symmetrical backward-wave coupler [1,2] as shown in Figure 10.5. The amount of phase shift with respect to a reference line depends on the phase of the reflection coefficient [1].

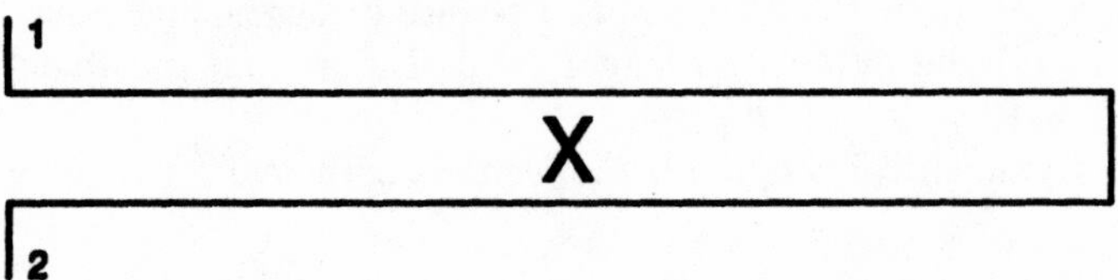

Figure 10.5. Nonuniform symmetrical backward-wave coupled line fixed phase shifter.

10.4 CONTINUOUSLY VARIABLE ANALOG PHASE SHIFTER

When a transmission line is terminated by a completely reactive circuit, the incident wave is reflected and the phase of the reflected wave becomes a function of the reactive termination. If the termination is a variable capacitance device, say, varactor diode, it can be tuned by means of a bias voltage, hence resulting in any desired phase shift [3–10].

A wideband, reflection-type phase shifter can be realized by the use of a −3 dB 90° directional coupler. Both backward-wave and forward-wave directional couplers can be used in the physical realization as shown in Figure 10.6. The operation of the hybrid coupled (−3 dB, 90° couplers are usually called *hybrids*) phase shifter is straightforward. The signals reflected from the identical diode terminations add up at the isolated port and cancel out at the input port (the performance critically depends on the isolation, diode terminations, coupling balance, and phase quadrature of the coupler). At each state of the diodes the signal exiting from the normally isolated port has a different phase, which is called the *transmission phase. Phase shift* is defined as

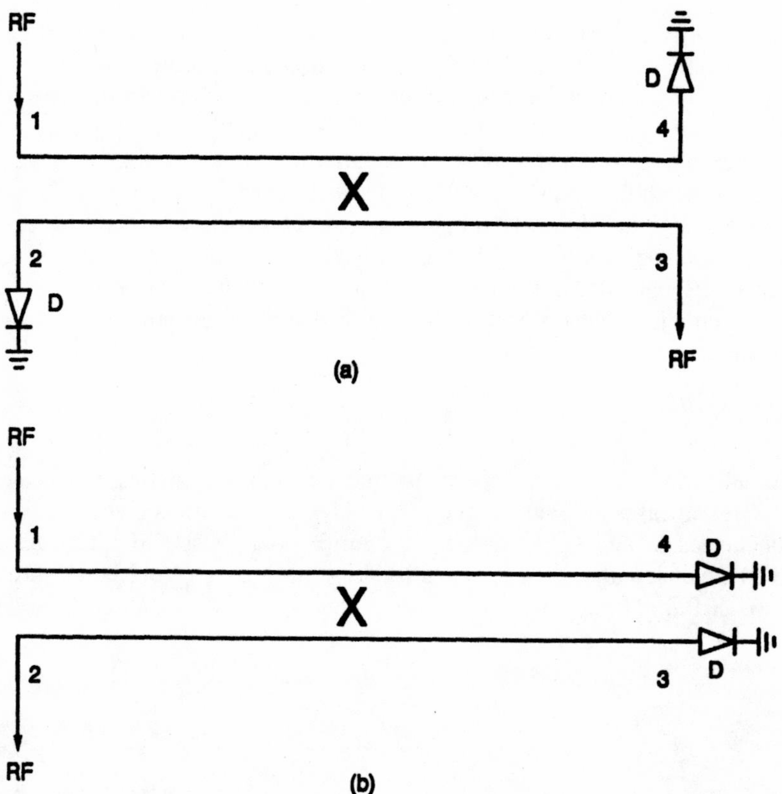

Figure 10.6 Analog reflection-type diode phase shifters using (a) −3 dB backward-wave coupler and (b) −3 dB codirectional coupler.

the change in transmission phase of the network. Thus, at two different states having respective transmission phases of ϕ_1 and ϕ_2, the phase-shift is

$$\Delta\phi = \phi_1 - \phi_2 \tag{10.2}$$

The sign of the phase shift depends upon the reference state.

The amount of phase shift achievable from a single circuit depends on the capacitance ratio of each of the identical diodes used in the terminations. The amount of capacitance ratio required for the same amount of phase shift in a given bandwidth can be obtained from the following relationship:

$$M = \frac{C_{\max}}{C_{\min}} = \left(\frac{f_{\max}}{f_{\min}}\right)^2 \tag{10.3}$$

where C_{min} and C_{max} are the diode capacitances at maximum and minimum bias voltage, respectively, and f_{max} and f_{min} are the upper and lower band edges.

Narrow, moderate, and ultrawideband -3 dB, 90° performance can be achieved from nonuniform directional couplers. Therefore, the achievable phase shift from a single-section phase shifter depends on the current diode technology. For an octave-bandwidth phase shifter the required M is 4, which is easily achievable. In this case, codirectional couplers (Figure 10.6(b)) are a better choice because the adjacent output ports enable the use of a single biasing network for the diode terminations.

For a 10:1 bandwidth, the required M is 100, which is not possible to achieve with the present diode technology. But an ultrawideband phase shifter can be designed with M_1 and M_2 defined as

$$M = M_1 M_2 \tag{10.4}$$

To maintain a flat phase shift versus frequency characteristics, it is preferred to use two different cascaded sections (i.e., $M_1 \neq M_2$). This circuit is given in Figure 10.7. The inductances L_1 and L_2 are needed to optimize the phase shift; they are usually realized as small lengths of bond wires (0.025-mm-diameter gold bond wire has an inductance of 0.8 nH per mm).

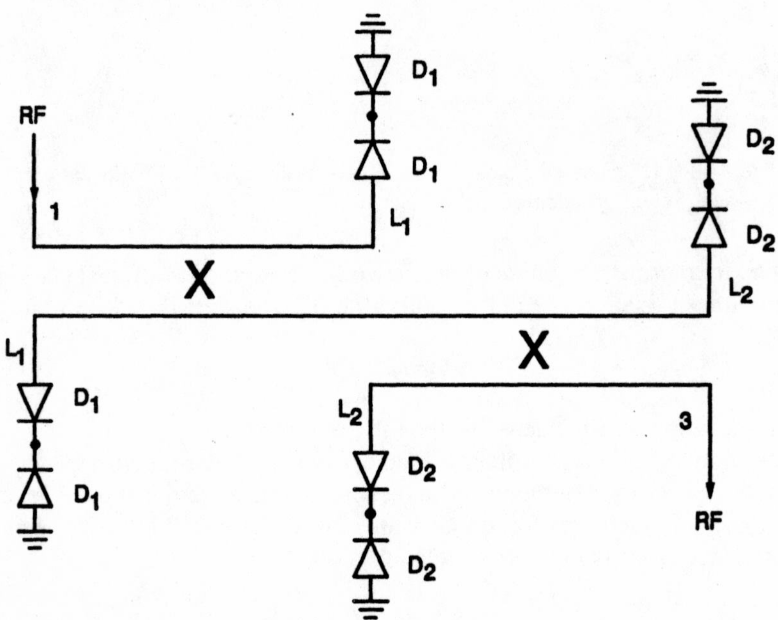

Figure 10.7 Continuously variable analog phase shifter with two nonidentical sections for an ultrawideband flat phase shift.

It is usually difficult to simultaneously obtain the required capacitance ratio and the minimum capacitance required in an ultrawideband phase-shifter design. Back-to-back connected diodes can provide the desired minimum capacitance without affecting the capacitance ratio. Therefore, the first section is responsible for the phase shift below the band center and the second section gives the desired phase shift above the band center.

The analysis of this circuit can be carried out by first considering each section on its own. Thus, for the first section,

$$\phi_1(V) = \pi - 2\tan^{-1}X_1(V) \tag{10.5}$$

and for the second section,

$$\phi_2(V) = \pi - 2\tan^{-1}X_2(V) \tag{10.6}$$

where $X_1(V)$ and $X_2(V)$ are the normalized reactances as function of bias voltage and are given as

$$X_1(V) = \frac{1}{Z_0}\left(\omega L_1 - \frac{1}{\omega C_1(V)}\right) \tag{10.7}$$

$$X_2(V) = \frac{1}{Z_0}\left(\omega L_2 - \frac{1}{\omega C_2(V)}\right) \tag{10.8}$$

where Z_0 is the characteristic impedance $C_1(V)$ and $C_2(V)$ are the bias voltage-dependent varactor diode capacitances.

For a given varactor diode, its voltage-dependent capacitance can be given as

$$C(V) = \frac{C_j(0)}{(1 + V/\psi_v)\gamma} \tag{10.9}$$

where $C_j(0)$ is the diode junction capacitance at $V = 0$ volts, ψ_v is the built-in potential, and γ is the capacitance-voltage slope exponent. The total transmission phase, $\phi(V)$, can be obtained as

$$\phi(V) = \phi_1(V) + \phi_2(V) \tag{10.10}$$

or

$$\phi(V) = 2\pi - 2\{\tan^{-1}X_1(V) + \tan^{-1}X_2(V)\} \tag{10.11}$$

The phase shift of this circuit is defined as

$$\Delta\phi(V) = \phi(V_{max}) - \phi(V) \tag{10.12}$$

If the maximum phase shift is defined as Φ, we have

$$\Phi = \phi(V_{max}) - \phi(V_{min}) \tag{10.13}$$

where V_{max} and V_{min} are the maximum and minimum bias voltages.

The maximum phase shift can also be written as

$$\Phi = 2\tan^{-1}p - 2\tan^{-1}q \tag{10.14}$$

where

$$p = \frac{X_1(V_{min}) + X_2(V_{min})}{1 - X_1(V_{min})X_2(V_{min})} \tag{10.15}$$

$$q = \frac{X_1(V_{max}) + X_2(V_{max})}{1 - X_1(V_{max})X_2(V_{max})} \tag{10.16}$$

For $\Phi = 180°$, we have

$$pq = -1 \tag{10.17}$$

For a desired maximum phase shift, Φ is optimized to obtain the required values for the inductors L_1 and L_2. For the optimization, the values of diode junction capacitances at $V = 0$ can be varied together with L_1 and L_2. When varying these component values, physical realizability (i.,e., the value of M) of the diodes should be kept in mind.

The physical implementation of the phase-shifter circuit shown in Figure 10.7 requires the design of a bias network for the diodes. The bias network can be realized by thin film resistors, which can be etched (or deposited) on the substrate (for example, tantalum with 50 Ω per sq. or 100 Ω per sq. can be used on alumina substrate). Grounding for diodes can be achieved by via holes. The constructional details for the ultrawideband phase shifter are shown in Figure 10.8.

10.5 MATCHING NETWORKS

Directional couplers can be used as wideband matching networks in circuits having highly reflective devices. The most common of such circuits is a balanced amplifier in

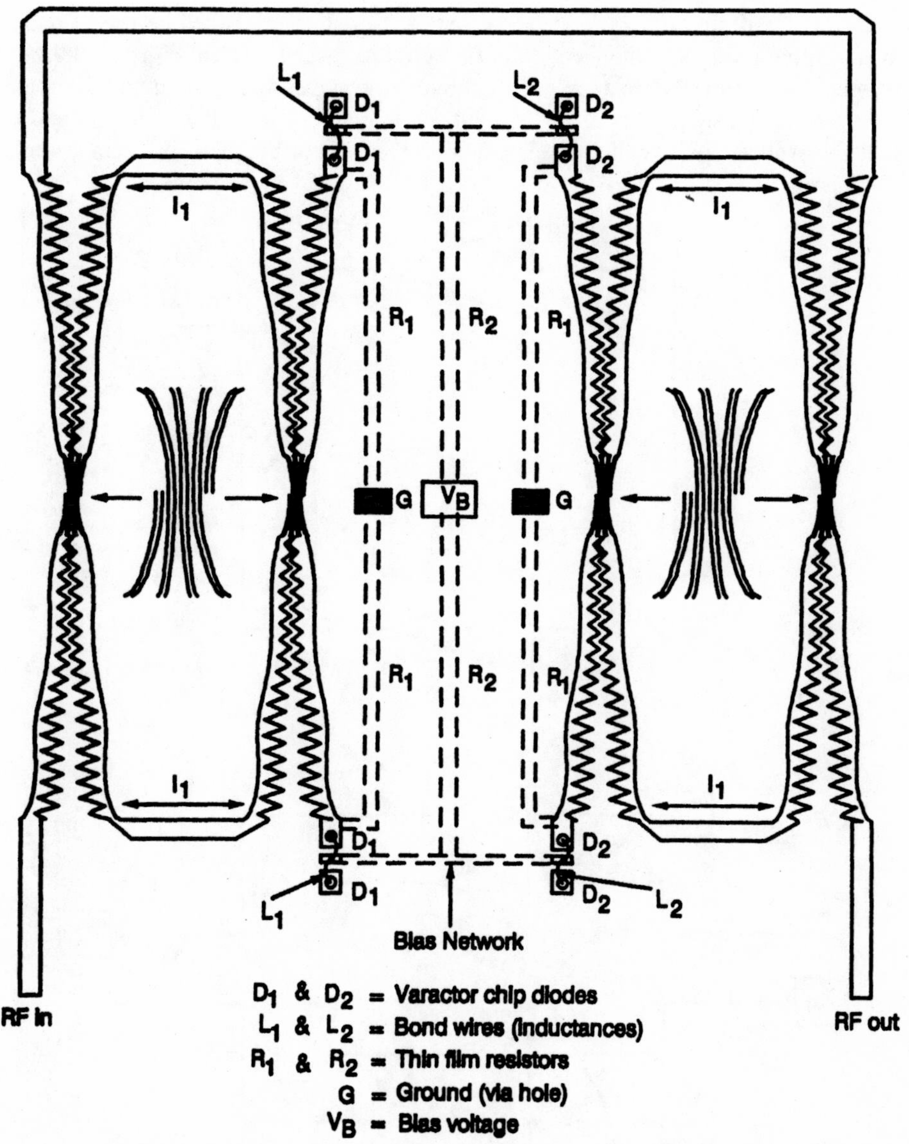

Figure 10.8 Two-stage, ultrawideband phase-shifter layout with bias network.

which input and output -3 dB quadrature directional couplers can be used to match the amplifier circuits. Balanced amplifier circuits are given in Figure 10.9. It is assumed that the amplifiers are identical, having the same input and output reflections. The two reflected signals entering the input coupler cancel out at the input port and add up at the isolated port, which is terminated. The output coupler performs a similar

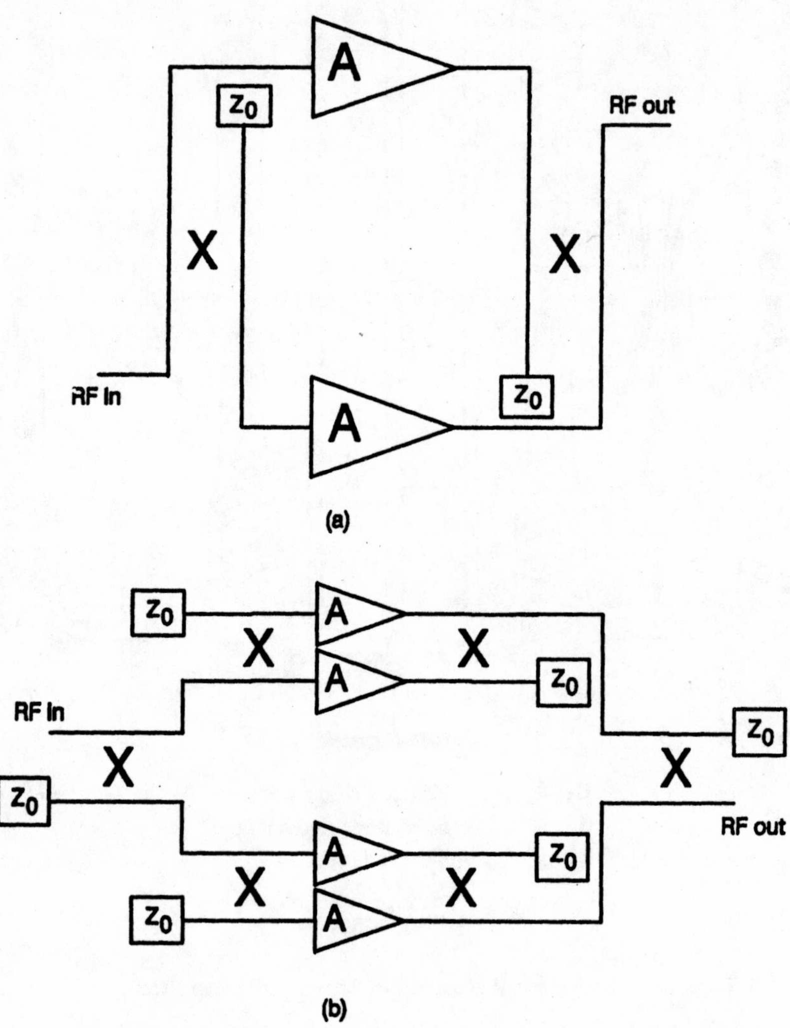

Figure 10.9. Some -3 dB quadrature couplers used as wideband matching networks in balanced amplifiers: (a) with double-coupled backward-wave couplers and (b) with codirectional couplers.

function. The balanced amplifier configuration [11,12] is very important because it is capable of delivering 3 dB higher output power compared to a single device. In amplifier circuits employing solid-state devices (FETs, MESFETs, HEMTs, BJTs) it is not possible to simply cascade several components to increase the output power. To achieve a high output power either the current or the voltage has to be increased. The maximum voltage and current that can be handled by a solid-state device depends on the device's internal characteristics and is usually limited by the breakdown (voltage) and burnout (current) characteristics of the devices.

10.6 BEAM-FORMING NETWORKS FOR PHASED ARRAY ANTENNAS

A phased array antenna has several antenna elements whose excitations are controlled by electronic means, thereby enabling the array to scan the beam in two or three dimensions without the need for mechanical rotation of the antenna. Phased array antennas can be designed in two ways: (1) a transmitting receiving (T/R) module for each antenna element or group of elements and (2) by dividing the array into identical subblocks. In the first approach the T/R module may include all the elements required for a radiator: driver, source, modulator, power amplifier, mixer, local oscillator, bandpass filter, T/R switch, low-noise amplifier, phase shifter, and antenna. The received signal is fed to a power combiner where a number of received signals can be combined for further processing. It is also possible to include the down-converter unit (mixer for IF extraction) and the detector unit in the same T/R module. However, the contents of a T/R module depends on the application and the targeted production cost with a given technology.

In the second approach, a number of antenna elements (subblocks) are fed from a common source. This arrangement requires a power divider-combiner network to divide the signal among the elements of the subarrays. Each antenna element is preceded by its own phase shifter and preferably a bandpass filter.

After this brief introduction to array antennas we shall attempt to identify the role of directional couplers in their design. However, before we do this, we should highlight one more point relevant to the first approach involving the design of array antennas using T/R modules. Power combining and division can be done either at RF or IF. Including a down-converter within the T/R module means that power combining takes place at IF. This effectively eliminates the need for RF directional couplers, which could be used in the power combiner/divider circuitry. On the other hand, the complexity and cost of a T/R module increases with the increasing number of components. Power combining at RF reduces the number of down-converters from hundreds (possibly thousands) to merely a few (number of required down-converters downline the RF combiner-divider circuits depends on the combined power level). Therefore, a compromise in the selection of type of circuitry can be decided by cost, design limitations, and performance expectations.

In microstrip antennas circular polarization can be achieved by using a -3 dB 90° directional coupler as the feed network, as illustrated in Figure 10.10. The antenna element is fed from two sides with equal signal amplitudes but with a 90° phase difference between them. Employing a square patch ensures radiation at the same frequency for both feeds. This antenna with its associated feed network can be accommodated in the same T/R module. Both bandpass-type and forward-wave directional couplers can be employed with any type of substrate since they do not require any bond wires.

There is a growing demand in the design of multifrequency antennas. In microwave communication and radar, most systems employ frequency diversity to enhance the reception quality of the transmitted signal. These systems require broadband antennas for both transmission and reception. If the antenna is not broadband enough the advantage of using frequency diversity is lost because the spectral separation of the transmission frequencies would be limited by the bandwidth of the antenna. Also, a broadband antenna is not a perfect radiator. Full, 100% coupling from an antenna to free space can take place only at fixed frequencies. Therefore, it is advantageous to excite a single antenna at two or more frequencies. The associated feed network of multifrequency antennas should be capable of operating at those frequencies. Such feed networks should therefore be broadband. However, the complexity of feed networks increases with increasing bandwidth. We propose two solutions: the use of (a) periodic couplers and (b) codirectional couplers.

Periodic couplers have the added advantage of filtering at the same time as feeding. The use of periodic couplers in a linear array is shown in Figure 10.11. The design of periodic couplers was given in Chapter 8. The feed network for the linear array can be analyzed by assuming that the array elements radiate at frequencies f_1 and

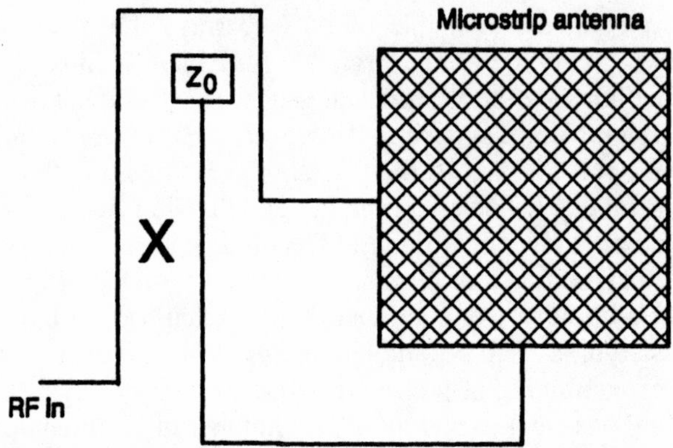

Figure 10.10. A -3 dB quadrature coupler feed network for circular polarization in microstrip antennas.

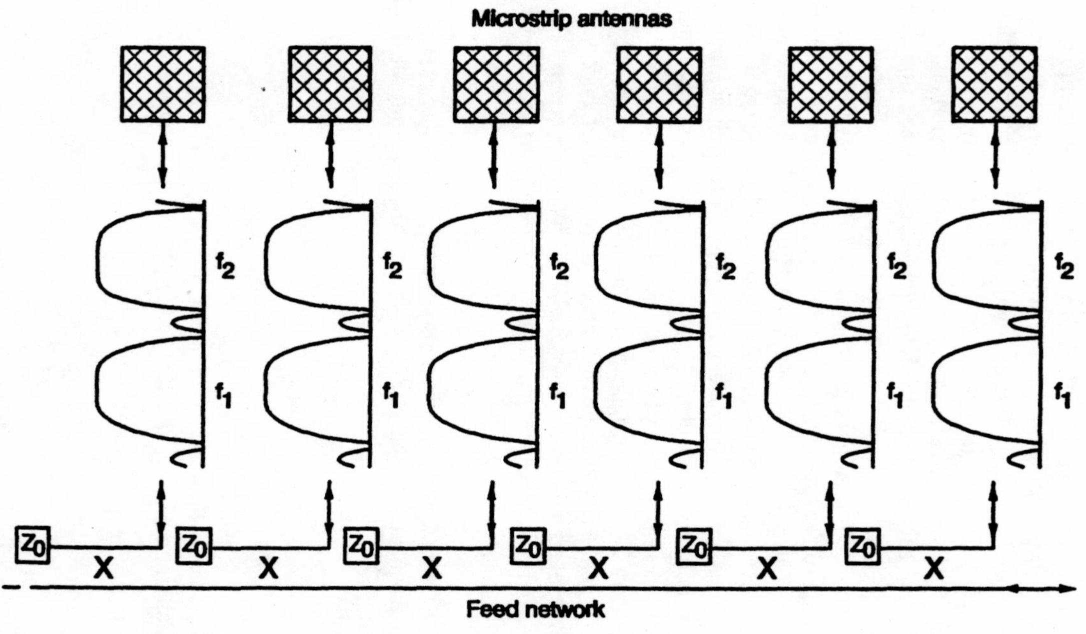

Figure 10.11. Periodic couplers used as beam-forming networks in multifrequency antenna arrays.

f_2. We shall also assume that f_1 and f_2 are separated by an appreciable amount (otherwise a single passband could do the job). We define two passbands for the directional coupler at f_1 and f_2. With a moderate length coupler, the out-of-band coupling is about 20 dB less than the passband coupling. This amount is adequate for most systems, provided that the noise level is less than the signal level. High levels of noise in the passband can mean service interruption, and its solution is the job of the coder-decoder or the design itself (that is why frequency diversity is used). The number of directional couplers in the feed network is decided mainly by the loss in the network. The design of the proceeding couplers (2, 3, 4, . . ., n) should ideally include the loss in the preceding couplers (see Figure 10.11). That is, the amount of coupling at f_1 and f_2 is increased by the amount of loss experienced by the signal. Because a wide range of coupling values can be achieved by periodic couplers employing double-coupled lines, an increase in coupling level does not constitute a major problem.

One of the most important applications of directional couplers is in a frequency-scanned phased array [13]. The simplest way of scanning a beam is by changing the frequency, thereby achieving a progressive phase shift in the feed network of the array. This requires a wideband feed network whose bandwidth is determined by the required frequency change to achieve the desired phase shift in order to scan the beam in a wide angle. A two-dimensional feed network utilizing this principle is illustrated in Figure 10.12.

Assume that the path length from the source to the first antenna element is l. At a given excitation frequency f_1 the phase of the signal appearing at the input of the antenna A is given by

$$\theta_1 = \frac{2\pi}{\lambda_1} l \tag{10.18}$$

The other elements have the same path lengths from the source, which gives the same phase shift at their inputs. At frequency f_2 the phase of the signal becomes

$$\theta_2 = \frac{2\pi}{\lambda_2} l \tag{10.19}$$

Because the direction of the beam is a function of this angle, the beam points at different directions for each phase of the input signal. The required frequency variation can be calculated for a desired scan angle. For example, if this angle is π, the required frequnecy variation is obtained by

$$\theta_2 - \theta_1 = \pi \tag{10.20}$$

where θ_1 and θ_2 are for the end-fire positions of the beam. Therefore,

$$\frac{1}{\lambda_2} - \frac{1}{\lambda_1} = \frac{1}{2l} \tag{10.21}$$

or

$$f_2 - f_1 = \frac{v}{2l} \tag{10.22}$$

where v is the velocity of propagation in the guide.

10.7 Σ–Δ CIRCUITS

The sum and difference of two signals can be formed by using backward-wave couplers, forward-wave couplers, or a combination of both. Some of the possible Σ–Δ configurations are given in Figure 10.13. The first type, shown in Figure 10.13(a), uses backward-wave couplers. The signal A is fed into an almost 0 dB coupler whose direct and isolated ports are terminated. For moderate to wideband applications, bandpass-type or periodic-type (for multiband operation) nonuniform directional couplers can be used. In these cases, the noisier signal is connected to the input of the 0 dB coupler because it also acts as a filter. The almost 0 dB coupler provides a 90° phase shift for signal A compared to a reference line whose length is equal to the coupler length. The signal B is connected to the isolated port of the −3 dB coupler whose input is connected to the output of the 0 dB coupler. The length of the reference line is equal to the total delay experienced by signal A at the input of the −3 dB coupler. The sum and difference of A and B appear at the output ports of the −3 dB coupler.

The second type, given in Figure 10.13(b), uses 0 dB and −3 dB codirectional couplers [14]. The operation of the circuit is similar to the first type. The usable bandwidth with moderate codirectional couplers is around 25%. Excellent performance ($\Delta < -25$ dB) is achieved in 15% bandwidth. Because codirectional couplers can be designed with no wiggling and very large gaps, they are most suitable for superconductive applications.

The third type uses almost 0 dB backward-wave and −3 dB codirectional couplers. This configuration is shown in Figure 10.13(c). This Σ–Δ is useful when filtering for one of the signals and adjacent Σ and Δ outputs are required.

The last circuit, given in Figure 10.13(d), uses 0 dB codirectional and −3 dB backward-wave couplers. This circuit can easily achieve 50% bandwidth with bandpass-type −3 dB couplers. Multiband operation is also possible by approximately trebling the length of the 0 dB codirectional and employing periodic-type −3 dB backward-wave couplers. The terminated isolated port of the 0 dB coupler can be used as the input for the Σ signal to obtain adjacent Σ and Δ ports, which also eliminates the need for terminations. However, in this case the operational bandwidth is affected by

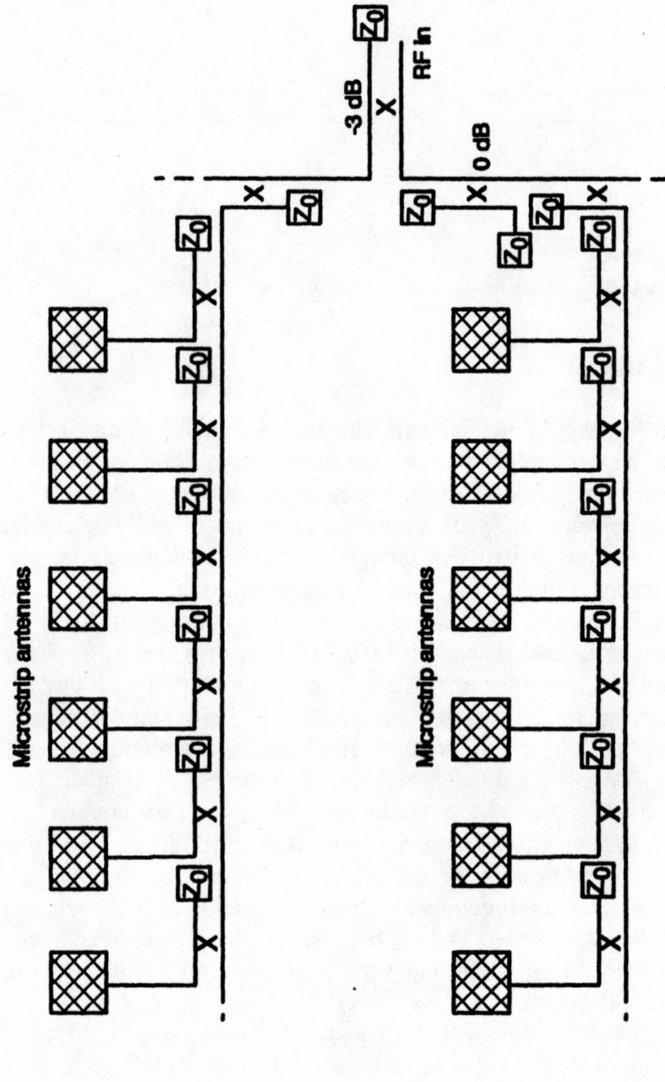

Figure 10.12. Codirectional couplers used as beam-forming networks in frequency-scanned phased array antennas.

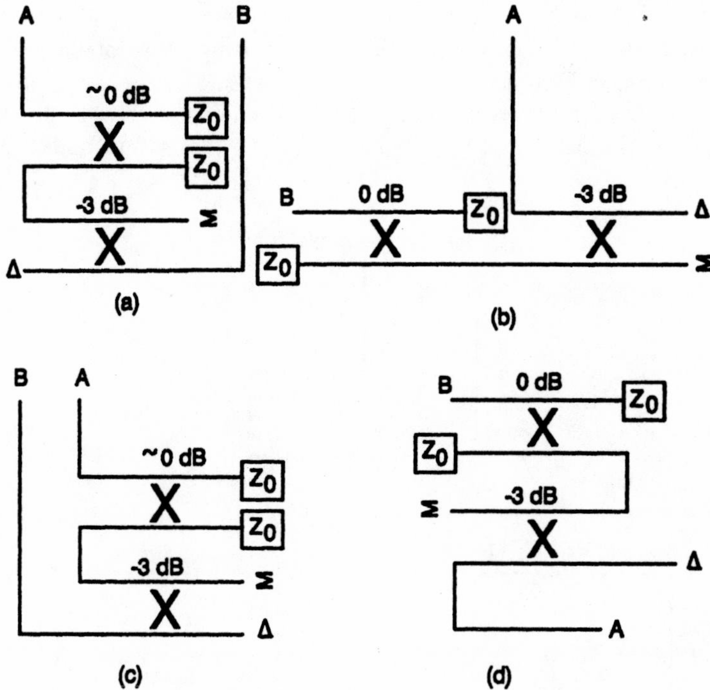

Figure 10.13. Microstrip Σ–Δ circuits using (a) backward-wave couplers, (b) codirectional couplers, (c) almost 0 dB backward-wave and −3 dB codirectional couplers, and (d) 0 dB codirectional and −3 dB backward-wave couplers.

the amount of uncoupled (direct) signal of the 0 dB coupler. Isolation in codirectional couplers can easily be maintained below −30 dB, which has negligible effect on the overall performance. However, for the best performance, the original circuit should be used wherever possible.

10.8 MONOPULSE COMPARATORS

Microstrip monopulse comparators are the key elements in some radar systems. They are used in the feed networks of phased array radar antennas to provide range, azimuth, and elevation information of a moving object. These systems are usually referred to as *Monopulse radars* because information about the target is obtained from the echo of a single pulse.

Monopulse comparators can be realized by any combination of Σ–Δ circuits, which are discussed in the previous section. A K_a-band microstrip monopulse comparator using -3 dB branch line couplers is reported by Jackson and Newman [15]. The general configuration of a monopulse comparator using four antenna elements (or subarrays) is given in Figure 10.14.

The terminated port gives the diagonal difference, which is not used. The operation of the comparator is reciprocal, and it can be used to both transmit and

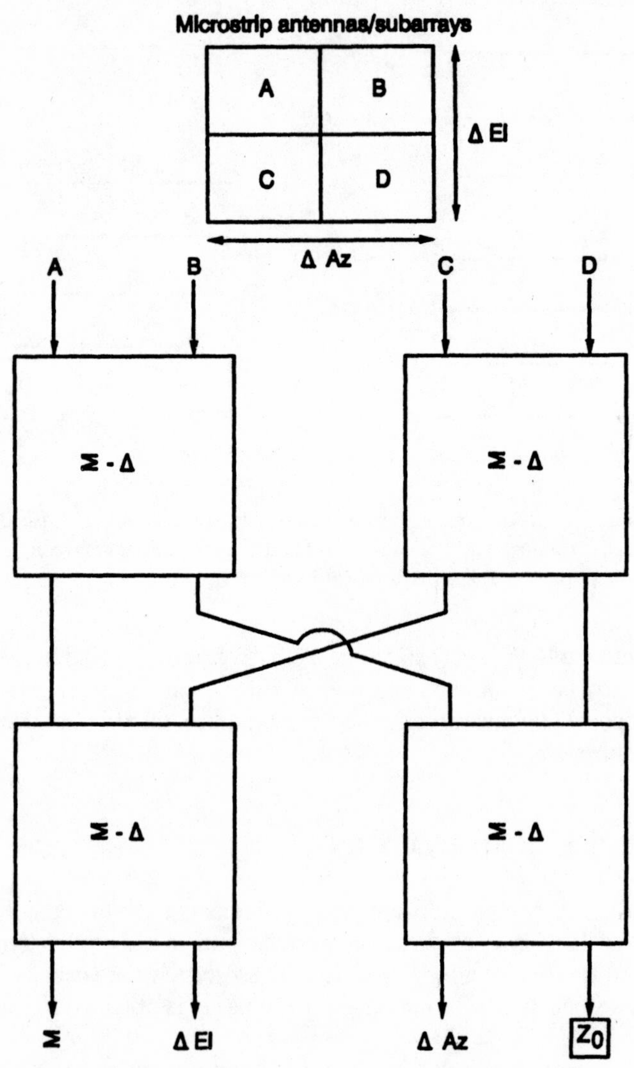

Figure 10.14. Microstrip monopulse comparator using Σ–Δ networks.

receive. The crossover can be avoided by modifying the configuration or by selecting the coupler type. Codirectional couplers can provide either a 90° or −90° phase shift, which gives significant design flexibility in the realization of comparators.

A monopulse comparator realized using forward-wave couplers only is given in Figure 10.15. The signals *B* and *D* are connected to 0 dB couplers that provide 90° phase-shift with respect to the signals *A* and *C*. A third 0 dB coupler is employed to provide for the sum of *A* and *B* and the difference of *C* and *D*. It also acts as a crossover for these signals. The Σ, ΔEl, and ΔAz signals are obtained with the indicated arrangement of *A, B, C,* and *D* antennas/subarrays. It should be noted here that the monopulse feed does not have to be on the same plane with the antennas; they can be designed on different surfaces using composite substrates sharing a common ground plane, as in capacitive or aperture-coupled microstrip antennas.

10.9 MIXERS

Microwave and millimeter wave technologies have not yet fully developed to deliver the high-quality signal processing equipment required in communication and radar systems. However, there are visible signs that the operational frequency of such equipment is steadily rising to microwave region.

Mixers [16] are used to up-convert an IF signal to a high frequency RF signal or down-convert an RF to an IF signal. The conversion from one frequnecy to another is done with the help of a local oscillator. Directional couplers play a major role in the design of mixers. They provide a balanced division of RF and LO signals for the mixer diodes.

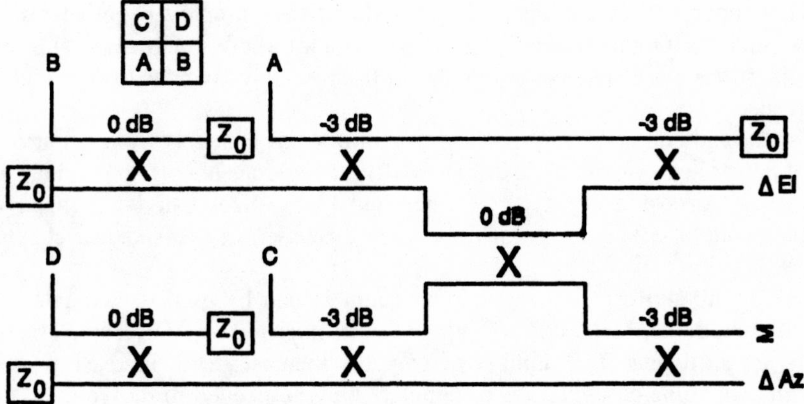

Figure 10.15. Microstrip monopulse feed network with forward-wave couplers.

Mixer circuits using -3 dB, 90° couplers are shown in Figure 10.16. The first type is a single diode mixer using an almost 0 dB bandpass backward-wave directional coupler. The LO signal is supplied from the isolated port and is not coupled. This arrangement provides some filtering for the RF signal. It also slightly increases the signal loss but additional filtering is not required.

The second type, which is given in Figure 10.16(b), is the most common single balanced mixer configuration. Image rejection is achieved by wiggly, single coupled line sections with $\lambda/4$ phase-shifting lines connected to the isolated ports. The -3 dB coupler in this filter is codirectional and provides in excess of 30 dB isolation for the RF and LO signals. In satellite communication, uplink and downlink use different frequencies to minimize interference between the channels. The satellite transponder employs a mixer to perform RF to RF frequency translation. Such a mixer can be realized by using either codirectional or bandpass-type couplers. Either type can provide significant bandwidth improvement in terms of coupling balance, phase quadrature, and isolation compared to ring or branch-line couplers (ring and branch-line couplers lack true phase quadrature because they possess dc continuity). An RF to RF mixer realized by again using codirectional couplers is shown in Figure 10.16(c).

10.10 OTHER APPLICATIONS

Directional couplers are extensively used in measuring equipment. They provide matching, dc isolation, sampling, and power division and combining. They can also be used in detector circuits. Construction of phase detectors are quite similar to mixers; in the case of amplitude detectors, diodes connected to the direct and coupled ports of a -3 dB coupler have the same orientation.

Another important application of directional couplers is in reflection-type amplifiers. Their operation is similar to phase shifters; they employ negative resistance devices (such as Gunn diodes) instead of varactor diodes. However, for proper operation, they require isolators or circulators because they give rise to very high input mismatches.

Directional couplers are also used to monitor power levels from a source. The sampled power level is usually less than 20 dB, which does not affect the main signal power level. Almost every type of filter and channelizer can be realized using directional couplers. The design of these were discussed in two separate chapters in this book.

Finally, nonuniform line directional couplers can be used to generate a wide variety of functions. For simple functions (as in microwave interferometers) a few couplers are sufficient. The number of couplers increases with increasing function complexity. In some cases, arrays of couplers may be needed to derive the required function.

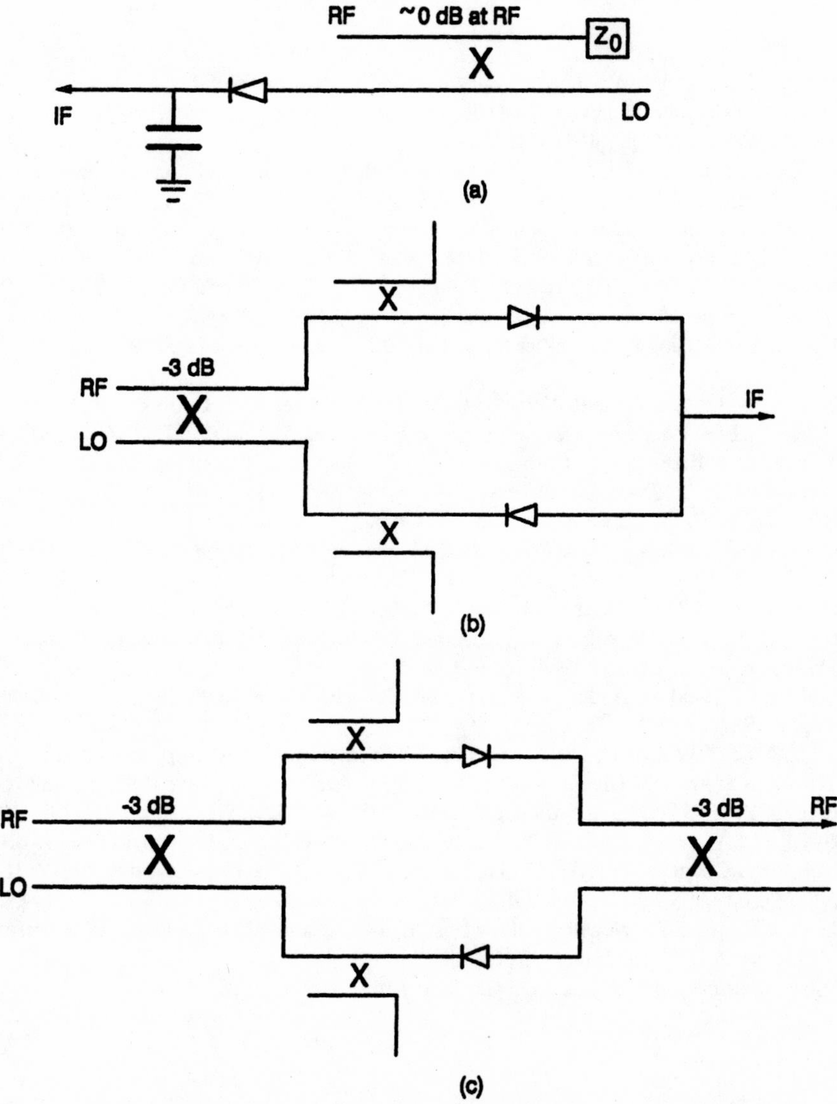

Figure 10.16. Diode mixer circuits using 90° directional couplers: (a) single-diode mixer, (b) balanced mixer, and (c) RF to RF frequency translator.

REFERENCES

[1] Shelton, J. P., and J. A. Mosko, "Synthesis and Design of Wide-Band Equal-Ripple TEM Directional Couplers and Fixed Phase Shifters," *IEEE Trans. on Microwave Theory and Tech.,* Vol. MTT-14, 1966, pp. 462–473.

[2] Schiek, B., and J. Kohler, "A Method for Broad-Band Matching Differential Phase Shifters," *IEEE Trans. on Microwave Theory and Tech.,* Vol. MTT-25, 1977, pp. 666–671.

[3] Henoch, B. T., and P. Tamm, "A 360° Reflection-Type Diode Phase Modulator," *IEEE Trans. on Microwave Theory and Tech.,* Vol. MTT-19, 1971, pp. 103–105.

[4] Starski, P. J., "Optimization of the Matching Network for a Hybrid Coupler Phase Shifter," *IEEE Trans. on Microwave Theory and Tech.,* Vol. MTT-25, 1977, pp. 662–666.

[5] Hopfer, S., "Analog Phase Shifter for 8–18 GHz,"*Microwave J.,* March 1979, pp. 48–50 and 84–85.

[6] Ulriksson, B., "Continuous Varactor-Diode Phase Shifter with Optimized Frequency Response," *IEEE Trans. on Microwave Theory and Tech.,* Vol. MTT-27, 1979, pp. 650–654.

[7] Yahara, T., Y. Kadowaki, H. Hoshika, and K. Shirahata, "Broad-Band 180° Phase Shift Section in *X*-Band," *IEEE Trans. on Microwave Theory and Tech.,* Vol. MTT-23, 1975, pp. 307–309.

[8] Niehenke, E. C., V. V. DiMarco, and A. Friedburg, "Linear Analog Hyperabrupt Varactor Diode Phase Shifters," *IEEE MTT-S Intl. Microwave Symp. Digest,* 1985, pp. 657–660.

[9] Dawson, D. E., A. L. Conti, S. H. Lee, G. F. Shade, and L. E. Dickens, "An Analog *X*-Band Phase Shifter," *IEEE Monolithic Circuits Symp. Digest,* 1984, pp. 6–10.

[10] Boire, D. C., J. E. Degenford, and M. Cohn, "A 4.5 to 18 GHz Phase Shifter," *IEEE Intl. Microwave Symp. Digest.,* 1985, pp. 601–604.

[11] Halladay, R., and K. Anderson, "Dual MMICs Deliver 1 W at K_u Band," *Microwave J.,* August 1987, pp. 168–178.

[12] MITEQ Inc., "Wideband, Ultra Low Noise GaAs FET Amplifiers Outperform Bipolar Counterparts in Traditional Bipolar Frequency Range," *Microwave J.,* July 1987, pp. 137–144.

[13] Hansen, R. C., *Microwave Scanning Antennas,* Vol. #2, New York: Academic Press, 1966.

[14] Uysal, S., J. Watkins, and C. W. Turner, "Sum-Difference Circuits Using 0 dB and −3 dB Co-Directional Couplers for Hybrid Microwave and MIMIC Circuit Applications," *IEEE MTT-S Intl. Microwave Symp. Digest,* Vol. 2, 1992, pp. 937–940.

[15] Jackson, C. M., and J. Newman, "Low Cost, K_a-band Microstrip Patch Monopulse Antenna," *Microwave J.,* July 1987, pp. 125–131.

[16] Maas, S., *Microwave Mixers,* Dedham, MA: Artech House, 1986.

Index

The Artech House Microwave Library

Acoustic Charge Transport: Device Technology and Applications, R. Miller, C. Nothnick, and D. Bailey

Algorithms for Computer-Aided Design of Linear Microwave Circuits, Stanislaw Rosloniec

Analysis, Design, and Applications of Fin Lines, Bharathi Bhat and Shiban K. Koul

Analysis Methods for Electromagnetic Wave Problems, Eikichi Yamashita, editor

Automated Smith Chart Software and User's Manual, Leonard M. Schwab

C/NL: Linear and Nonlinear Microwave Circuit Analysis and Optimization Software and User's Manual, Stephen A. Maas

Capacitance, Inductance, and Crosstalk Analysis, Charles S. Walker

Design of Impedance-Matching Networks for RF and Microwave Amplifiers, Pieter L. D. Abrie

Digital Microwave Receivers, James B. Tsui

Electric Filters, Martin Hasler and Jacques Neirynck

E-Plane Integrated Circuits, P. Bhartia and P. Pramanick, editors

Feedback Maximization, Boris J. Lurie

Filters with Helical and Folded Helical Resonators, Peter Vizmuller

GaAs FET Principles and Technology, J. V. DiLorenzo and D. D. Khandelwal, editors

GaAs MESFET Circuit Design, Robert A. Soares, editor

GASMAP: Gallium Arsenide Model Analysis Program, J. Michael Golio, et al.

Handbook of Microwave Integrated Circuits, Reinmut K. Hoffmann

Handbook for the Mechanical Tolerancing of Waveguide Components, W.B.W. Alison

HEMTs and HBTs: Devices, Fabrication, and Circuits, Fazal Ali, Aditya Gupta, and Inder Bahl, editors

High-Power Microwave Sources, Victor Granatstein and Igor Alexeff, edtiors

High-Power GaAs FET Amplifiers, John Walker, editor

High-Power Microwaves, James Benford and John Swegle

Introduction to Microwaves, Fred E. Gardiol

Introduction to Computer Methods for Microwave Circuit Analysis and Design, Janusz A. Dobrowolski

Introduction to the Uniform Geometrical Theory of Diffraction, D. A. McNamara, C. W. I. Pistorius, and J. A. G. Malherbe

LOSLIN: Lossy Line Calculation Software and User's Manual, Fred E. Gardiol

Lossy Transmission Lines, Fred E. Gardiol

Low-Angle Microwave Propagation: Physics and Modeling, Adolf Giger

Low Phase Noise Microwave Oscillator Design, Robert G. Rogers

MATCHNET: Microwave Matching Networks Synthesis, Stephen V. Sussman-Fort

Matrix Parameters for Multiconductor Transmission Lines: Software and User's Manual, A. R. Djordjevic, et al.

MIC and MMIC Amplifier and Oscillator Circuit Design, Allen Sweet

Microelectronic Reliability, Volume I: Reliability, Test, and Diagnostics, Edward B. Hakim, editor

Microelectronic Reliability, Volume II: Integrity Assessment and Assurance, Emiliano Pollino, editor

Microwave and RF Circuits: Analysis, Synthesis, and Design, Max Medley

Microwave and RF Component and Subsystem Manufacturing Technology, Heriot-Watt University

Microwave Circulator Design, Douglas K. Linkhart

Microwave Engineers' Handbook, 2 Volumes, Theodore Saad, editor

Microwave Materials and Fabrication Techniques, Second Edition, Thomas S. Laverghetta

Microwave MESFETs and HEMTs, J. Michael Golio, et al.

Microwave and Millimeter Wave Heterostructure Transistors and Applicatons, F. Ali, editor

Microwave and Millimeter Wave Phase Shifters, Volume I: Dielectric and Ferrite Phase Shifters, S. Koul and B. Bhat

Microwave and Millimeter Wave Phase Shifters, Volume II: Semiconductor and Delay Line Phase Shifters, S. Koul and B. Bhat

Microwave Mixers, Second Edition, Stephen Maas

Microwave Transmission Design Data, Theodore Moreno

Microwave Transition Design, Jamal S. Izadian and Shahin M. Izadian

Microwave Transmission Line Couplers, J. A. G. Malherbe

Microwave Tubes, A. S. Gilmour, Jr.

Microwaves: Industrial, Scientific, and Medical Applications, J. Thuery

Microwaves Made Simple: Principles and Applicatons, Stephen W. Cheung, Frederick H. Levien, et al.

MMIC Design: GaAs FETs and HEMTs, Peter H. Ladbrooke

Modern GaAs Processing Techniques, Ralph Williams

Modern Microwave Measurements and Techniques, Thomas S. Laverghetta

Monolithic Microwave Integrated Circuits: Technology and Design, Ravender Goyal, et al.

Nonlinear Microwave Circuits, Stephen A. Maas

Nonuniform Line Microstrip Directional Couplers, Sener Uysal

Optical Control of Microwave Devices, Rainee N. Simons

PLL: Linear Phase-Locked Loop Control Systems Analysis Software and User's Manual, Eric L. Unruh

Scattering Parameters of Microwave Networks with Multiconductor Transmission Lines: Software and User's Manual, A.R. Djordjevic, et al.

Solid-State Microwave Power Oscillator Design, Eric Holzman and Ralston Robertson

Terrestrial Digital Microwave Communications, Ferdo Ivanek, et al.

Time-Domain Response of Multiconductor Transmission Lines: Software and User's Manual, A.R. Djordjevic, et al.

Transmission Line Design Handbook, Brian C. Waddell

Yield and Reliability in Microwave Circuit and System Design, Michael Meehan and John Purviance